Lecture Notes in Mathematics

Editors:
J.-M. Morel, Cachan
F. Takens, Groningen
B. Teissier, Paris

Marc Bernot · Vicent Caselles
Jean-Michel Morel

Optimal Transportation Networks

Models and Theory

 Springer

Authors

Marc Bernot
Unité de mathématiques pures et appliquées
ENS Lyon
46, allée d'Italie
69363 Lyon Cedex 7, France
mbernot@umpa.ens-lyon.fr

Jean-Michel Morel
CMLA, Ecole Normale Supérieure de
Cachan
61 av. du Président Wilson
94235 Cachan Cedex, France
morel@cmla.ens-cachan.fr

Vicent Caselles
Dept. de Tecnologies de la Informació
i les Comunicacions
Pompeu Fabra University
Passeig de Circumval.lació 8
08003 Barcelona, Spain
vicent.caselles@upf.edu

ISBN: 978-3-540-69314-7   e-ISBN: 978-3-540-69315-4
DOI: 10.1007/978-3-540-69315-4

Lecture Notes in Mathematics ISSN print edition: 0075-8434
                    ISSN electronic edition: 1617-9692

Library of Congress Control Number: 2008931162

Mathematics Subject Classification (2000): 49Q10, 90B10, 90B06, 90B20

Cover design: SPi Publishing Services

Printed on acid-free paper

9 8 7 6 5 4 3 2 1

springer.com

Preface

The transportation problem can be formalized as the problem of finding the optimal paths to transport a measure μ^+ onto a measure μ^- with the same mass. In contrast with the Monge-Kantorovich formalization, recent approaches model the branched structure of such supply networks by an energy functional whose essential feature is to favor wide roads. Given a flow φ in a tube or a road or a wire, the transportation cost per unit length is supposed to be proportional to φ^α with $0 < \alpha < 1$. For the Monge-Kantorovich energy, $\alpha = 1$ so that it is equivalent to have two roads with flow $1/2$ or a larger one with flow 1. If instead $0 < \alpha < 1$, a road with flow $\varphi_1 + \varphi_2$ is preferable to two individual roads φ_1 and φ_2 because $(\varphi_1 + \varphi_2)^\alpha < \varphi_1^\alpha + \varphi_2^\alpha$. Thus, this very simple model intuitively leads to branched transportation structures. Such a branched structure is observable in ground transportation networks, in draining and irrigation systems, in electric power supply systems and in natural objects like the blood vessels or the trees. When $\alpha > 1 - \frac{1}{N}$ such structures can irrigate a whole bounded open set of $\mathbb{R}^N$.

The aim of this set of lectures is to give a mathematical proof of several existence, structure and regularity properties empirically observed in transportation networks. This will be done in a simple mathematical framework (measures on the set of paths) unifying several different approaches and results due to Brancolini, Buttazzo, Devillanova, Maddalena, Pratelli, Santambrogio, Solimini, Stepanov, Xia and the authors.

The link with anterior discrete physical models of irrigation and erosion models in hydrography and with discrete telecommunication and transportation models will be discussed. It will be proved that most of these models fit in the simple model sketched above. Several mathematical conjectures and questions on the numerical simulation will be developed.

The authors thank Bernard Sapoval for introducing them to this subject and for giving them many insights on physical aspects of irrigation networks. V. Caselles acknowledges partial support by the "Departament d'Universitats, Recerca i Societat de la Informació de la Generalitat de Catalunya" and by PNPGC project, reference BFM2003-02125. J.M.Morel acknowledges many discussions with and helpful suggestions from Giuseppe Devillanova, Franco Maddalena, Filippo Santambrogio and Sergio Solimini. He also thanks UCLA for its hospitality during the revision of the manuscript.

Table of Contents

1 Introduction: The Models

The aim of this book is to give a unified mathematical theory of branched transportation (or irrigation) networks. The only axiom of the theory is a $l \times s^\alpha$ cost law ($0 \leq \alpha \leq 1$) for transporting a good with size s on a path with length l. Let us explain first why this assumption is relevant.

Humans have designed many supply-demand distribution networks transporting goods from supply sites to widespread distribution sites. This is obviously the case with networks for ground transportation, communication [44], electric power supply, water distribution, drainage [52], or gas pipelines [16]. These networks show a striking similarity to observable natural irrigation and draining systems which connect a finite size volume to a source or to an outlet. Forests, plants, weeds, and trees together with their root systems, but also the nervous, the bronchial and the cardiovascular systems have a common morphology which seems to derive from topological constraints together with energy saving requirements. All of these systems look like spatial trees and succeed in spreading out a fluid from a volume or a source onto another volume. The associated morphology is a tree (or a union of trees) made of bifurcating vessels. Their intuitive explanation is that transport energy is saved and better protection is obtained by using broad vessels as long as possible rather than thin spread out vessels.

The above list involves a huge range of natural and artificial phenomena. So the underlying optimization problems have been treated at several complexity levels, by different communities, and with different goals. Even the names of the network optimization problems vary. The problem first emerged in the framework of operational research and graph theory. In that case the geometry of the network is fixed *a priori* and the problem is known in graph theory by the name of *Minimum Concave Cost Flows*. This model was introduced in Zangwill's article [100]. It considers graphs endowed with flows and prescribed sources and demands at certain graph nodes. The transportation of a mass along an edge has a cost, usually proportional to its length but *concave with respect to the mass*. The optimization problem is to find the minimal cost flow achieving a prescribed transport. The abundant literature dealing with this problem refers to all classical operational research applications: transportation, communication, network design and distribution, production and inventory planning, facility/plant location, scheduling and air traffic control.

M. Bernot et al., *Optimal Transportation Networks*. Lecture Notes in Mathematics 1955.
© Springer-Verlag Berlin Heidelberg 2009

Fig. 1.1. The structure of the nerves of a leaf (see [97] for a model of leaves based on optimal irrigation transport).

In all of these cases, the concavity of the transport cost with respect to the flow along an edge is justified by economical arguments. In Zangwill's words: *The literature is replete with analyses of minimum cost flows in networks for which the cost of shipping from node to node is a linear function. However, the linear cost assumption is often not realistic. Situations in which there is a set-up with charge, discounting, or efficiencies of scale give rise to concave functions.*

A similar view is developed in the more recent article [99]: *Although a mathematical model with a linear arc cost function is easier to solve, it may not reflect the actual transportation cost in real operations. In practice, the unit cost for transporting freight usually decreases as the amount of freight increases. The cargo transportation cost in particular is mainly influenced by the cargo type, the loading/unloading activities, the transportation distance, and the amount. In general, each transportation unit cost decreases as the amount of cargo increases, due to economy of scale in practice. Hence, in actual operations the transportation cost function can usually be formulated as a concave cost function.*

Regarding the practical resolution of this optimization problem, the key source of complexity of the concave cost network flow problem arises from the minimization of a concave function over a convex feasible region, defined by the network constraints. As Zangwill points out [100]: *although concave functions can be minimized by an exhaustive search of all the extreme points of the convex feasible region, such an approach is impractical for all but the simplest of problems.* Indeed, there are potentially an enormous number of local optima in the search space. For this reason, concave cost network flow problems are known to be NP-hard [46] and [51]. Yet, many algorithms have been developed over the years for solving these problems [45], [46], [85], [41], [99] (this last reference thoroughly discusses the differences between existing algorithms). As an argument towards a complexity reduction, Zangwill proves in [100] that optimal flows have a tree structure. Indeed, a local optimum is necessarily an extremal point so that two flow paths connecting two points have to coincide. We shall prove the very same result, under the name of single path property, in a more general framework (see Chapter 7).

In a similar context many authors have considered the problem of optimizing branched distribution such as rural irrigation, reclaimed water distribution, and effluent disposal (see [31], [35], [77] among numerous other references). This literature is more specialized and the model is made more precise: the layout is prescribed and the decision variables include design parameters (pipe diameters, pump capacities, and reservoir elevations). The objective function to be minimized reflects the overall cost construction plus maintenance costs. The constraints are in the form of demands to be met and pressures at selected nodes in the network to be within specified limits.

A big limitation of Minimum Concave Cost Flows is that the geometry of the network is fixed. In practice the network itself has to be designed! This problem cannot really be addressed within graph theory and leads us to more geometrical considerations. In a very recent paper [53], Lejano developed a method for determining an optimal layout for a branched distribution system *given only the spatial distribution of potential customers and their respective demands*. Lejano insists on the novelty of such a problem with respect to Minimum Concave Cost Flows: *Much research has been developed around optimizing pipeline design assuming a predetermined geographical layout of the distribution system. There has been less work done, however, on the problem of optimizing the configuration of the network itself. Generally, engineers develop the basic layout through experience and sheer intuition.*

In this book, we shall retain only the essential aspect of the above problems, namely the concavity of the transport cost. The optimization problem we shall consider is the more general one, namely the optimization of the layout itself. Since the number of sources and wells is finite, most irrigation or transport practical models look at first discrete. Yet, because of the huge scale ratios and of invariance requirements, a continuous model is preferable.

In the continuous framework, the transportation or irrigation problem can be formalized as the problem of finding the optimal paths to transport any positive measure (not necessarily atomic or finite) μ^+ onto another positive measure μ^- with the same mass. The first and classical statement of this transportation problem is due to Monge [62] and its formalization to Kantorovich [50]. In this original model, masses are transported on infinitely many straight routes by infinitesimal amounts. Probably the best natural phenomenon akin to this mathematical solution is the nectar gathering by bees from the fields to the hive. In the Monge-Kantorovich model the straightness of trajectories makes the transportation cost $\varphi \times l$ strictly proportional to the amounts of transported goods φ and to the distance l. As we already pointed out, transportation on straight lines is neither economically nor energetically sound in most situations and does not correspond to the observed morphology of transportation networks.

Fig. 1.2. A cast of a dog set of lungs. They solve the problem of bringing the air entering the trachea onto a surface with very large area (about 500 m^2 for the human lungs). A mathematical and physical study of the lungs efficiency is developed in [61].

The simplest mathematical model compatible with the above considerations on supply networks uses a cost function whose essential feature is to favor wide routes rather than thin ones. Given a flow φ in a road or a tube or a wire, the transportation cost per unit length is taken proportional to φ^α with $0 < \alpha < 1$. For the Monge-Kantorovich energy $\alpha = 1$ so that it is equivalent to have two roads with flow $1/2$ or a larger one with flow 1. If instead $0 < \alpha < 1$, a road with flow $\varphi_1 + \varphi_2$ is preferable to two individual roads with flows φ_1 and φ_2 because $(\varphi_1 + \varphi_2)^\alpha < \varphi_1^\alpha + \varphi_2^\alpha$. In the terms of Xia [94], the interpretation reads as follows: *In shipping two items from nearby cities to the same far away city, it may be less expensive to first bring them into a common location and put them on a single truck for most of the transport. In this case, a "Y shaped" path is preferable to "V shaped" path.* So the only axiom in this theory will be the s^α cost law with $0 \le \alpha \le 1$.

To the best of our knowledge the concave power law was first proposed in 1967 by Gilbert [44] to optimize communication networks. But the very same model has recurred in the past twenty years for the physical analysis of scaling laws in animal metabolism [88], [89], [8]. The models in the mentioned authors and in the river basins literature described extensively in the book [73] consider finite graphs G made of tubes satisfying the Kirchhoff conservation law. The source and wells are modeled by finite sums of Dirac masses. The energy of the network, interpretable as a power dissipation, is

$$W(G) = \sum_{k \in G} l_k s_k^{-2} \varphi_k^2,$$

Fig. 1.3. A very old tree (1200 years) spans his branches towards the light. Trees and plants solve the problem of spanning their branches as much as possible in order to maximize the amount of light their leaves receive for photosynthesis. The surface of the branches is minimized for a better resistance to parasites, temperature changes, etc.

where G is the set of tubes, k the tube index, s_k the tube section, l_k the tube length, and φ_k the flow in the tube. Thus the model is *a priori* more complex than the Gilbert model, which only considers a flow depending cost

$$\tilde{W}(G) = \sum_k l_k \varphi_k^\alpha \qquad (1.1)$$

However, it will be proven in Chapter 14 that the energy W reduces to a Gilbert energy $\tilde{W}$ with $\alpha = \frac{2}{3}$, under the mild and natural assumption that the network volume is fixed.

Let us give some examples. A tree (see Figure 1.3), a plant or a forest can be viewed as transportation networks from the ground (a 3-D volume) onto a 2D surface, typically a sphere in the case of an isolated tree or, at a different scale, a plane in the case of a forest. Indeed, the roots spread in the ground to attain every part of the underneath volume in search of water and nutriments. In the air, branches tend to spread out to intercept as much sunlight as possible. Thus we can roughly view the branches as means to reach by subdivision a sphere approximating the tree's foliage. The tree branches are barked bundles of fibers going from the ground to the leaves (see Figure 1.1) and transporting the sap at constant speed. Up to a multiplicative constant, the flux in a branch is equal to its section: $\varphi_k = s_k$ where s_k is the section. The obvious protection requirement of the branches from external aggressions such as parasites or temperature changes leads to minimize their area, which is barked for the same reason. Thus we are led to a minimal surface problem,

$$\min_{G}(\tilde{W}(G) = \sum_{k} l_{k}\varphi_{k}^{\frac{1}{2}}) \tag{1.2}$$

with the constraints that the graph G satisfies the Kirchhoff law and irrigates a stipulated volume and a stipulated surface. The cost given in (1.2) is similar to the one for pipe-lines [16]. In that case the construction cost is $W(G) = \sum_{k} l_{k}\varphi_{k}^{\frac{1}{2}}$ because it is proportional to the length and to the diameter of the tubes while the flow is itself roughly proportional to the size of the tubes. The discrete model is well justified in that latter case since wells and plants are indeed finite atomic masses. In the case of O.C.N. (Optimal Channel Networks, [73]) modeling river networks, the power is again $\alpha \simeq \frac{1}{2}$ and the irrigation constraints have to model the drainage of a whole region to a few river mouths. Last but not least, the human body contains irrigation networks irrigating a whole volume from the heart in the case of arteries (see Figure 1.4) and a very large surface from the throat in the case of lungs (see Figure 1.2).

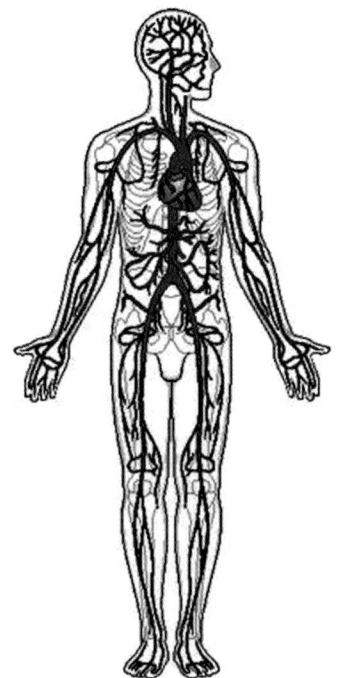

Fig. 1.4. Arteries of the human body. They solve the problem of transporting the blood from the heart to the whole body with very low basal metabolic rate. Attempts to demonstrate scaling laws in Nature have focused on the basal metabolic rate [89, 90]. This rate has been linked to the total blood flow.

All of these networks have a striking similarity structure. In particular, they are trees.

The Gilbert energy is simple but the devil is in the irrigation constraints which we just mentioned. In the discrete model these are just a finite set. Of course all irrigated Lebesgue measures can be approximated by finite atomic measures. Yet in the continuous setting, the feasibility of irrigation networks is no more granted. Not that there would be a geometric obstruction to the existence of infinite trees irrigating a positive volume K in a strong sense, namely with a branch of the tree (a sequence of tubes) arriving at <u>every</u> point of K. Such tube trees can be constructed by rather explicit rules; they can satisfy the Kirchhoff law and can even have the fluid speed decrease and be null at the tips of capillaries. Such constructions can be found (e.g.) in [8], [67] and [27]. Figure 1.5 gives an intuitive recursive construction of a tree irrigating a cube. One of the first examples described in the literature is due to Besicovitch [15] and is precisely the construction in Figure 1.5.

Optimal irrigation networks are complex objects and some generic descriptors are needed to describe them. Among the candidates, fractal dimensions have had the preference of most authors. In geomorphology, an early study of the fractal-like behavior of natural drainage networks was started as early as 1945 by R.E. Horton [47], A.N. Strahler [82], and generalized by E. Tokunaga [84].

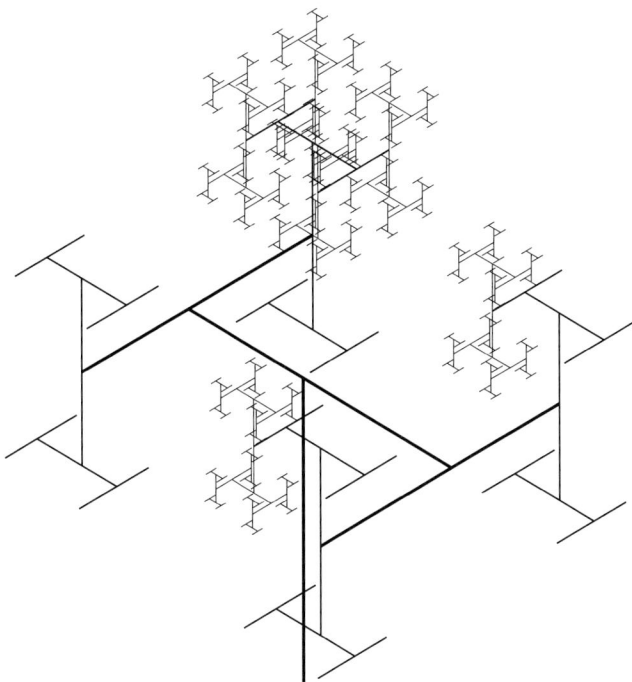

Fig. 1.5. An irrigating tree.

To be able to derive fractal dimensions from the variational model, a basic assumption is usually made, namely that the network has a branched tree structure made at each scale of tubes of a certain uniform length, radius and with a given branching number. In other terms, the irrigation system is a fully homogeneous tree in scales, sizes and shapes like the tree of Figure 1.5. Then, under these *ad hoc* assumptions, the authors of [88], [89], [20] prove that the network has a fractal structure with self-similar properties. The irrigation network is then characterized by the branching ratios and the ratios of radii and lengths of the tubes. Calling n the branching ratio, the above assumptions permit to conclude the radii and length ratios scale as powers of n. This heuristic reasoning ends up with a structure described as a self-similar fractal.

The weak point of the above treatment is the very strong homogeneity assumption involved in the heuristic calculations. The self-similarity properties, if they are really true, should be deduced from first principles. The basic variational principle related to the cost of irrigation and the irrigation constraints should be the only basis for structural statements as was requested in [88]. Authors in [89] acknowledge that *In spite of the very large number of numerical and empirical studies, no general theory based on fundamental laws has yet been developed for (...) fractal behavior (...).*

Thus, the aim of these lecture notes is to go back on the foundations of optimal channel networks. We shall define a common mathematical structure to all of them and give a proof of several structure and regularity properties empirically observed in transportation networks. These results hopefully pave the way to the study of fractal properties. They already confirm that fully self-similar models are too simplistic. Actually many of the questions raised by specialists such as the regularity issues, the existence and shape of river basins, the dimensionality of the irrigated volumes, surfaces and of the irrigation network itself can only be rigorously treated once a continuous invariant model has been stated and existence and regularity proven in the general setting adopted here. For example the formation of "tubes", which is one of the assumptions of the empirical models must (and will) be demonstrated, and their regularity and branching number estimated.

The first job is to fix an adequate mathematical object for irrigation networks, simple enough to cope with the mathematical challenges but general enough to cover the variety of cases. At least three mathematical models have been proposed so far. Xia [94] modeled the networks as currents which can be approximated in the sense of currents by finite irrigation networks. This definition based on a relaxation yields easy weak existence results for minimizers but does not seem well adapted to a thorough description of the network structure.

Maddalena and Solimini [59] gave a much more explicit definition in the case of networks which start from a single source. Their definition is directly derived from the tree model as a union of fibers. The fibers are Lipschitz paths $t \mapsto \chi(\omega, t)$ depending upon a parameter. Branches are defined as sets of fibers which stick together. Such fiber trees were called *patterns*.

The authors of the present volume extended this Lagrangian framework to more general finite and infinite graphs. Let G be finite graph with flows φ_k satisfying the Kirchhoff law at all vertices which are not sources or sinks. Then the flow on the graph can be written as the sum of a finite set of flows φ_i on distinct paths γ_i, all starting at a source vertex and ending at a sink vertex. Thus we can write the graph with its flow as an atomic finite measure on the set of all paths, namely $G = \sum_i \varphi_i \delta_{\gamma_i}$. This decomposition is classical in graph theory (see e.g. [49]). It leads to define in whole generality irrigation networks as *measures on the set of paths*. Such objects will be called *traffic plans*. This name takes into account the fact that not only the general circuit is described by a measure on the set of paths, but in fact all trajectories. It is indeed proven in Chapter 3 that a traffic plan can be parameterized as a set of paths or fibers $t \mapsto \chi(\omega, t)$, exactly like in the pattern model. This leads to a double representation of networks, either as a set of trajectories or as a measure on a set of paths. Typically, structure and regularity theorems will involve the parametric representation, while the existence and compactness results are easier in the measure on paths framework. Both formalisms make it very easy to formalize the irrigation and Kirchhoff constraints.

Once the basic formal object is fixed, the next tasks on the agenda are to explore the existence of traffic plans joining arbitrary positive measures, and, if they exist, to find the structure of optimal traffic plans. We can roughly say that many properties, which in the above quoted physical explorations are taken as axioms, can be either proved, or disproved or in general made precise. Optimal networks will be proven to have a tree structure (no cycle) and to be in some sense a countable union of tubes. Even if the self-similar perfect structure assumed by some authors is clearly disproved (see Chapter 13), several regularity properties will be proven which open the way to the definition and exploration of fractal dimensions for optimal networks. We hope that these lectures will open more mathematic problems than they solve. Chapter 15 lists some of them.

2 The Mathematical Models

We shall review the four main formalisms proposed for transportation networks in a discrete and continuous framework. Section 2.1 is dedicated to the Monge-Kantorovich model. Section 2.2 describes the Gilbert-Steiner discrete irrigation model. Section 2.3 is devoted to the notation of the three continuous mathematical models (transport paths, patterns and traffic plans). Section 2.4.1 lists the mathematical questions and tells where they will be solved in the book. Section 2.5 discusses several extensions and related models in urban optimization.

2.1 The Monge-Kantorovich Problem

The seminal Monge problem [62] was to move a pile of sand from a place to another with minimal effort. In the Kantorovich [50] formalization of this problem, μ^+ and μ^- are positive measures on $\mathbb{R}^N$ which model respectively the supply and demand mass distributions. A transport scheme from μ^+ onto μ^- is described by telling where each piece of supplied mass is sent. In the Kantorovich formalism this information is encoded in a positive measure π on $\mathbb{R}^N \times \mathbb{R}^N$ where $\pi(A \times B)$ represents the amount of mass going from A to B. This measure π will be called a *transference plan* (also called *transport plan* in the literature). To evaluate the efficiency of a transference plan, a cost function $c : \mathbb{R}^N \times \mathbb{R}^N \to \mathbb{R}^+$ is considered where $c(x, y)$ is the cost of transporting a unit mass from x to y. The cost associated with a transference plan is $\int_{\mathbb{R}^N \times \mathbb{R}^N} c(x, y) d\pi(x, y)$. The minimization of this functional is the Monge-Kantorovich problem. This problem has attracted many mathematicians and is extremely well documented, as apparent in the recent and excellent treatises and reviews [2, 37, 38, 42, 86].

As an example, consider the cost function $c(x, y) = |x-y|^2$, and the supply and demand measures $\mu^+ = \delta_x$ and $\mu^- = \frac{1}{2}(\delta_{y_1} + \delta_{y_2})$, where δ_x stands for the Dirac mass at x. The minimizer π is the measure on $\mathbb{R}^N \times \mathbb{R}^N$ such that $\pi(\{x\} \times \{y_1\}) = \frac{1}{2}$ and $\pi(\{x\} \times \{y_2\}) = \frac{1}{2}$. The optimal transportation is achieved along geodesics between x, y_1 and y_2 as represented in Figure 2.1.

M. Bernot et al., *Optimal Transportation Networks*. Lecture Notes in Mathematics 1955.
© Springer-Verlag Berlin Heidelberg 2009

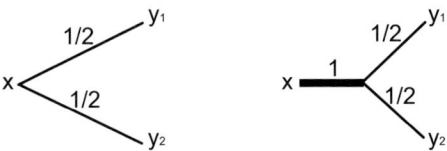

Fig. 2.1. The transport from δ_x to $\frac{1}{2}(\delta_{y_1} + \delta_{y_2})$. Monge-Kantorovich straight solution (left) versus Gilbert's branching one (right).

2.2 The Gilbert-Steiner Problem

In the Monge-Kantorovich framework, the cost of the structure achieving the transport is not modeled. Indeed, with this formulation, the cost behaves as if every single particle of sand went straight from its starting to its ending point. In the case of real supply-demand distribution problems, achieving this kind of single particle transport would be very costly. Thus from the economical viewpoint the Monge-Kantorovich problem is unrealistic. In most transportation networks, the aggregation of particles on common routes is preferable to individual straight ones. The local structure of human-designed distribution systems doesn't look as a set of straight wires but rather like a tree.

The Steiner problem which consists in minimizing the total length of a network connecting a given set of points could be such a transportation model. Yet this cost is not realistic because it does not discriminate the cost of high or low capacity edges (a road has not the same cost as a highway). The first model taking into account capacities of edges was proposed by Gilbert [44] in the case of communication networks. This author models the network as a graph such that each edge e is associated with a flow (or capacity) $\varphi(e)$. Let $f(\varphi)$ denote the cost per unit length of an edge with flow (or capacity) φ. It is assumed that $f(\varphi)$ is subadditive and increasing, i.e., $f(\varphi_1)+f(\varphi_2) \geq f(\varphi_1+\varphi_2) \geq \max(f(\varphi_1), f(\varphi_2))$. Gilbert then considers the problem of minimizing the cost of networks supporting a given set of flows between terminals. The subadditivity of the cost f translates the fact that it is more advantageous to transport flows together. Thus, it leads to delay bifurcations. In the fluid mechanics context, this subadditivity follows from Poiseuille's law, according to which the resistance of a tube increases when it gets thinner (we refer to [12, 27] for a study of irrigation trees in this context). The simplest model of this kind is to take $f(\varphi) = \varphi^\alpha$ with $0 < \alpha < 1$.

Following [44], consider atomic sources $\mu^+ = \sum_{i=1}^{k} a_i \delta_{x_i}$ and sinks $\mu^- = \sum_{j=1}^{l} b_j \delta_{y_j}$ with $\sum_i a_i = \sum_j b_j$, $a_i, b_j \geq 0$. An irrigation graph G is a weighted directed graph with straight edges $E(G)$ and a flow function $w : E(G) \rightarrow (0, \infty)$ satisfying Kirchhoff's law. Observe that G can be written as a vector measure

$$G = \sum_{e \in E(G)} \varphi(e) \mathcal{H}^1|_e \mathbf{e} \tag{2.1}$$

where $\mathbf{e}$ denotes the unit vector in the direction of e and $\mathcal{H}^1$ the Hausdorff one-dimensional measure. We say that G irrigates (μ^+, μ^-) if its distributional derivative ∂G satisfies

$$\partial G = \mu^- - \mu^+. \tag{2.2}$$

The Gilbert energy of G is defined by

$$M^\alpha(G) = \sum_{e \in E(G)} \varphi(e)^\alpha \mathcal{H}^1(e). \tag{2.3}$$

We call the problem of minimizing $M^\alpha(G)$ among all finite graphs irrigating (μ^+, μ^-) the Gilbert-Steiner problem. The Monge-Kantorovich model corresponds to the limit case $\alpha = 1$ and the Steiner problem to $\alpha = 0$. The structure of the minimizer of (2.3) with $\mu^+ = \delta_x$ and $\mu^- = \frac{1}{2}(\delta_{y_1} + \delta_{y_2})$ is shown in Figure 2.1. Let us finally mention that the Gilbert model is adopted in [16,52] to study optimal pipeline and drainage networks. From a numerical point of view, a backtrack algorithm exploring relevant Steiner topologies is proposed in [98] to solve a problem of water treatment network. A different algorithmic approach can be found in [101].

2.3 Three Continuous Extensions of the Gilbert-Steiner Problem

As we mentioned in the preface, the discrete Gilbert-Steiner model was only recently set in the Kantorovich continuous framework [94], [59] where the wells and sources are arbitrary measures, instead of finite sums of Dirac masses.

2.3.1 Xia's Transport Paths

Let μ^+, μ^- be two positive Radon measures with equal mass in a compact convex set $X \subset \mathbb{R}^N$. A vector measure T on X with values in $\mathbb{R}^N$ is called by Xia [94] a *transport path* from μ^+ to μ^- if there exist two sequences μ_i^-, μ_i^+ of finite atomic measures with equal mass and a sequence of finite graphs G_i irrigating (μ_i^+, μ_i^-) such that $\mu_i^+ \to \mu^+$, $\mu_i^- \to \mu^-$ weakly as measures and $G_i \to T$ as vector measures. The energy of T is defined by

$$M^\alpha(T) := \inf \liminf_{i \to \infty} M^\alpha(G_i), \tag{2.4}$$

where the infimum is taken over the set of all possible approximating sequences $\{\mu_i^+, \mu_i^-, G_i\}$ to T. Denote

$$M^\alpha(\mu^+, \mu^-) := \inf\{M^\alpha(T) : T \text{ is a transport path from } \mu^+ \text{ to } \mu^-\}.$$

If $\alpha \in (1 - \frac{1}{N}, 1]$, by Theorem 3.1 in [94], the above infimum is attained and finite for any pair (μ^+, μ^-). Xia shows in a series of papers several structure and regularity properties of optimal transport paths which we shall comment later on.

2.3.2 Maddalena-Solimini's Patterns

The Gilbert problem was given by Maddalena and Solimini [59] a different (Lagrangian) formulation in the case of a single source supply $\mu^+ = \delta_S$. The authors model the transportation network as a "pattern", or set of "fibers" $\chi(\omega, \cdot)$, where $\chi(\omega, t)$ represents the location of a particle $\omega \in \Omega$ at time t. The set Ω is an abstract probability space indexing all fibers. Without loss of generality one can take $\Omega = [0, 1]$ endowed with the Lebesgue measure on borelians E denoted by $\lambda(E) = |E|$. All the fibers are required to stop at some time $T(\omega)$ and satisfy $\chi(\omega, 0) = S$ for all ω, which means that all fibers start at the same source S. The set of fibers is given a structure corresponding to the intuitive notion of branches. Two fibers ω and ω' belong to the same branch at time t if $\chi(\omega, s) = \chi(\omega', s)$ for all $s \leq t$. Then the partition of Ω given by the branches at time t yields a time filtration. The branch of ω at time t is denoted by $[\omega]_t$ and its measure by $|[\omega]_t|$. The energy of the pattern is defined by

$$\tilde{E}^\alpha(\chi) = \int_\Omega \int_0^{T(\omega)} |[\omega]_t|^{\alpha-1} dt d\omega. \tag{2.5}$$

It is easily checked that this definition extends the Gilbert energy (2.3) when the discrete graph is a finite tree. The measure irrigated by the Maddalena et al. pattern is defined for every set A as the measure in Ω of the set of fibers stopping in A, $\mu^-(A) = |\{\omega : \chi(\omega, T(\omega)) \in A\}|$. Both the formulation of the energy and the irrigation constraints are very handy in this model and we shall retain its essential features. We can illustrate the Lagrangian formalism with the simplest non trivial example of transportation from one Dirac mass to two Dirac masses (see Figure 2.1). Maddalena-Solimini's solution is given by the set of fibers $\chi : [0, 1] \times [0, \infty) \to \mathbb{R}^2$, where $\chi(\omega, t)$ is either the path from x to y_1 (if $\omega \in [0, 1/2]$), or the path from x to y_2 (if $\omega \in (1/2, 1]$). This solution can also be written as the sum of two paths with flow $\frac{1}{2}$ each. Thus the solution is formalized either as a parameterized set of fibers or as the sum of two Dirac masses on the set of paths. This later point of view leads to the notion of *traffic plans* as developed in the next section.

2.3.3 Traffic Plans

In [11] the authors of the present book extended the pattern formalism to the case where the source is any Radon measure. They called "traffic plan" any positive measure on the set of Lipschitz paths (see Chapter 3 for a more accurate definition). This model allows one to add *who goes where* constraints to the optimization problem. Thus, it also gives a framework to the *mailing problem*.

The who goes where (or mailing) problem has the same energy but a different boundary condition. Indeed, it specifies which part of μ^+ goes to a given part of μ^-, instead of just requiring μ^+ to go globally onto μ^-. Thus it

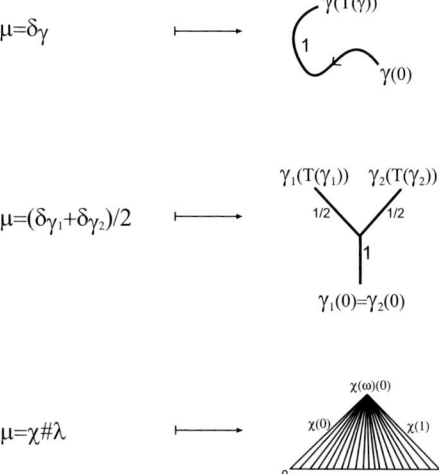

Fig. 2.2. Three traffic plans and their associated embedding : a Dirac measure on γ, a tree with one bifurcation, a spread tree irrigating Lebesgue's measure on the segment $[0,1] \times \{0\}$ of the plane. In the bottom example, to $\omega \in [0,1]$ corresponds $\chi(\omega) \in K$, the straight path from $(1/2, 1)$ to $(\omega, 0)$.

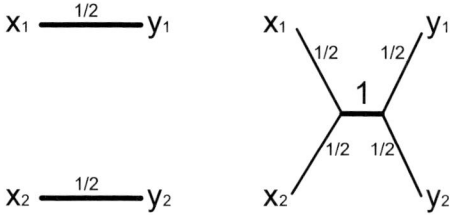

Fig. 2.3. Irrigation problem minimizer versus traffic plan minimizer with who goes where constraint. The energy is the same and the irrigating and irrigated measures are the same. In the second case, the who goes where constraint prescribes that the mass in x_1 must go y_2 and the mass in x_2 to y_1. The solution is quite different.

incorporates a *transference plan constraint* which we shall call the "who goes where" constraint. For example, consider $\mu^+ = \delta_{x_1} + \delta_{x_2}$ and $\mu^- = \delta_{y_1} + \delta_{y_2}$ as in Figure 2.3. If we want to find the best transportation network with the constraint that all the mass in x_1 is sent to y_2, and all the mass in x_2 is sent to y_1, we need another formalism, since the graph approach does not convey this kind of information. The Lagrangian formalism introduced in [59] is adapted to that problem, since it keeps track of every mass particle.

Figure 2.2 shows three examples of traffic plans: a Dirac mass on a finite length path γ (which means that a unit mass is transported from $\gamma(0)$ to $\gamma(L)$), a traffic plan with "Y" shape, and a traffic plan transporting a Dirac mass to the Lebesgue measure on a segment of the plane. As we mentioned in the preface, any graph with a flow satisfying Kirchhoff's law can be modeled by a finite atomic measure on the space of paths in the graph. Classical graph theory ensures that one can decompose the graph into a sum of paths starting on sources and ending on wells.

2.4 Questions and Answers

Thus, we have three different formalisms having as common denominator a cost per flow s proportional to s^α: the original discrete Gilbert model defined on finite graphs, a generalization by Xia's *transport path* model and a generalization by the traffic plan (measure on the set of paths) model. The irrigation pattern model [59] is a particular instance of traffic plan. Of course the rather intuitive equivalence of the continuous models and their consistency with the original Gilbert model will have to be checked. To be more precise:

1. Consistency problem: when the irrigated measures μ^+ and μ^- are finite atomic, do the minimizers for all continuous models coincide with the Gilbert minimizers?
2. For two general positive measures μ^+ and μ^- with equal mass, are the Xia's minimizers also optimal traffic plans and conversely?
3. When $\mu^+ = \delta_S$ is a single source, are optimal patterns and optimal traffic plans equivalent notions?

The answer to questions 1, 2 and 3 will be an easy consequence of a main structure property of optimal traffic plans: they have a "single path" structure, that is, contain at most one path joining any two points, as observable in most irrigation networks. In the case of optimal traffic plans for the irrigation problem, this result is easily extended to show that the network is a tree. This result was proven in the case of finite atomic measures by Gilbert. It is shown, again for discrete graphs in [94], and for irrigation patterns in [28,79]. The equivalence of all models was proven in [58] and [13] and is the object of Chapter 9.

Do optimal networks look like (infinite) graphs and what is their structure? A first natural question is whether optimal irrigation networks are countably rectifiable. This property has been proven in [96] for transport paths, in [59] for irrigation trees and in [11] for traffic plans. The next question towards a graph structure is to control the branching number. Gilbert [44] proved that there is a universal bound for the branching number of any minimal finite graph, as observable in Figure 1.1. We shall show the same property in general following [95] and [13]. Another strong result, proven in various forms in [13, 29, 95] is that any optimal irrigation traffic plan can be monotonically approximated by finite irrigation trees.

A main question raised by Xia [96] is finally the interior regularity, according to which the structure of an irrigation tree away from the supports of μ^+ and μ^- is a finite graph with straight edges and bounded branching number. In this book, we shall prove this regularity result for traffic plans when $\alpha > 1 - \frac{1}{N}$, using the results and techniques of [13,95]. As a consequence of the equivalence of models mentioned above, all regularity results apply to all models as well.

Let us mention further graph structure properties of optimal traffic plans. Following [95] we shall prove (Proposition 7.16) that any path in the irrigation graph with flow larger than a constant is bi-Lipschitz, with explicit estimates on the Lipschitz constant depending on α and the minimal value of the flow. Another feature of optimal networks shown in [95] is the following. For any point x of the optimal irrigation network and for any $\varepsilon > 0$, there is a ball $B(x, r)$ such that the network inside $B(x, r)$ can be additively decomposed in two parts : a main network P made of a finite number of bi-Lipschitz graphs and a residual network R whose total flow is less than ε. In addition, all points in the ball where the flow is larger than ε are contained in the support of P. Actually, following [13, 28, 78], this kind of result will be made global by a pruning Lemma (see Proposition 7.14).

2.4.1 Plan

Chapter 3 is devoted to all detailed definitions of the main terms used throughout the book: traffic plan and its parameterization, fiber, irrigated and irrigating measure, transference plan, stopping time of each fiber, the multiplicity (or flow) at each point and finally the energy. General existence theorems are given for the existence of optimal traffic plans with prescribed irrigated and irrigating measure or with prescribed transference plan under the mere assumption that a feasible solution exists.

Chapter 4 gives a more geometric form to traffic plans and their energy. Its first important result is that traffic plans with finite energy are countably rectifiable. This first regularity result is used to obtain the consistency of the traffic plan energy E^α with the original Gilbert energy. This consistency reads

$$E^\alpha(P) = \int_{\mathbb{R}^N} |x|_\chi^\alpha \, d\mathcal{H}^1(x) \qquad (2.6)$$

for loop-free traffic plans, where $|x|_\chi$ denotes the measure of the set of fibers passing by x.

Chapter 5 is devoted to a series of elementary operations permitting to combine traffic plans into new ones. New traffic plans can be built by restriction, concatenation, union, and hierarchical concatenation. This last operation gives a standard way to explicitly construct infinite irrigation trees, or patterns.

Chapter 6 uses these techniques to prove by explicit constructions that for $\alpha > 1 - \frac{1}{N}$ where N is the dimension of the ambient space, the optimal cost to transport μ^+ to μ^- is finite. Thus the energy of a minimal traffic plan between μ^+ and μ^-, denoted $E^\alpha(\mu^+, \mu^-)$ turns out to define a metric on the space of probability measures. This distance will be compared with the classical Wasserstein distance $E^1(\mu^+, \mu^-)$. These results will be generalized in Chapter 10 which defines, following Devillanova and Solimini [78], an *irrigability dimension* and compares it with classical Hausdorff and Minkowski dimensions. These results imply in particular that the Lebesgue measure of a cube in $\mathbb{R}^N$ is *not irrigable* for $\alpha \leq 1 - \frac{1}{N}$.

Chapter 7 proves several crucial structure properties for optimal traffic plans. First, the *single path property*, which implies that optimal traffic plan for the irrigation problem are trees. Second, a *monotone approximation of optimal traffic plans by finite graphs* irrigating atomic measures is constructed. The main results of this chapter have been proved in different contexts in [13, 58, 70, 79, 94, 95]. Our presentation follows [13].

Chapter 8 goes further and proves the finite graph structure of the irrigation network away from the irrigated measures. This is what Xia [96] calls "interior regularity". Finally a "boundary regularity", comes out, namely a universal bound on the number of branches at each point of the network. The main results of this chapter have been proved in different contexts in [13, 58, 94, 95] and again our presentation follows [13].

Chapter 9 uses the regularity results to prove the equivalence of all mentioned models: their minimizers are identical and their minima equal.

The four last chapters establish crucial links with the former discrete and physical theories and give several applications.

Chapter 11 which follows the work of Santambrogio [75] establishes a crucial equivalence of the theory with the theory of Optimal Channel Networks. It proves that the link established in this discrete theory between optimal irrigation and optimal landscape is still valid in the continuous framework. There is for every optimal pattern (traffic plan starting from a point source) a unique landscape function $z(x)$ up to an additive constant. Then the fibers of the optimal traffic plan turn out to be the steepest descent curves of the landscape. Here one sees one of the advantages of the continuous framework, which permits to ask and answer regularity issues. It is indeed proven that the optimal landscape is Hölder. Figure 2.4 shows various results of landscape optimization: the resulting river network is a tree with regular branches.

Chapter 12 investigates particular examples considering only discrete measures and deriving the optimal configurations in the simplest case: one source and two wells. This study yields a geometric procedure to construct optimal configurations when the data are not too large.

Using these techniques, Chapter 13 investigates the structure of an optimal traffic plan irrigating the Lebesgue measure on a segment from one source. It demonstrates that the irrigation tree is infinite and gives a geometric procedure and heuristics to construct a global optimum.

Chapter 14 proves that all irrigation problems fit into the framework given by the Gilbert energy. Starting from the "naive" embedded tube models

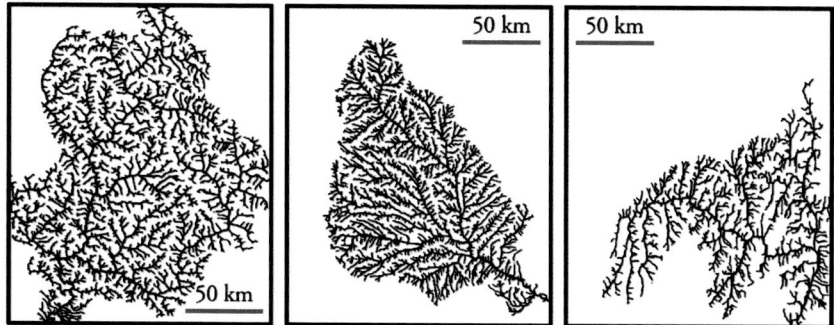

Fig. 2.4. Different river networks. These branched networks manage to collect the water raining over the whole territory they cover. According to [73], such a network evolves towards a locally minimal flow energy configuration.

and writing their physical Poiseuille energy, a simple Lagrangian argument shows that we are led back to the Gilbert energy. The main objection to the infinitesimal models is that Poiseuille law in 3D leads to $\alpha = \frac{2}{3} = 1 - \frac{1}{3}$ which is precisely the critical exponent, for which a Lebesgue measure can't be irrigated.

Finally Chapter 15 proposes a (certainly non-exhaustive) list of open problems.

2.5 Related Problems and Models

2.5.1 Measures on Sets of Paths

Although the name of traffic plan is new in the literature, there are several antecedents for the use of measures on sets of paths for transportation models. The idea of a probabilistic representation of sets of paths appears in many contexts (particularly for equations of diffusion type). The first reference in the context of conservation laws and fluid mechanics seems to be in Brenier [19], where it is used to describe particle trajectories for the incompressible Euler equation. Paolini and Stepanov [70] use normal one-dimensional currents but point out that they can be represented by measures over the metric set of Lipschitz paths in $\mathbb{R}^N$. Buttazzo, Pratelli and Stepanov [23] [22] used measures on paths in urban transportation models. They proposed to call them *transports*. Now this term along with the term *transport plan* and the term *transference plan* are being used with a different meaning in the context of the Monge-Kantorovich problem. Hence our choice of a new term for the object denoting a set of trajectories. Measures on set of paths are familiar in optimal transport theory, where transport maps and transference plans can be thought of in a natural way as measures in the space of minimizing geodesics [71]. See [7] for a similar approach within Mather's theory. In a more general context, [4] and [3] use measures in the space of continuous maps to characterize a flow with BV vector field and prove its stability. The Lecture Notes [87] contain, among several other things, a comprehensive treatment of the topic of measures in the space of action-minimizing curves, including at the same time the optimal transport and the dynamical systems case. This unified treatment was inspired by [9].

2.5.2 Urban Transportation Models with more than One Transportation Means

A series of papers by Buttazzo, Brancolini, Pratelli, Santambrogio, Solimini, and Stepanov has considered recently urban transportation models where two means for transportation with different cost structure or more compete.

The seminal papers seem to be [17], [24], [25] and [21], where a rather natural extension of the Monge-Kantorovich problem is being proposed to model urban transportation network. Given an initial and a final measure representing for example the homes and work places of citizens, it is assumed that the citizens dispose of two transportation means, namely a high speed and cheap network (typically the underground or highway network, see Figures 2.5 and 2.6). This network is modeled as a connected rectifiable set with finite length. Another low speed pedestrian network permits to access

Fig. 2.5. An old map of France's main roads. In contrast with the tree structure of hydrographic maps, the road network is a general network containing cycles. This network approximately solves another problem, the mailing, or "who goes where" problem. For this problem existence and structure results will be given in the traffic plan framework. However, these results are by far less complete than the results on the irrigation problem because of the lack of a tree structure. In particular the regularity theory is fully open (see Chapter 15 on the open problems).
Image from http://www.france-property-and-information.com/

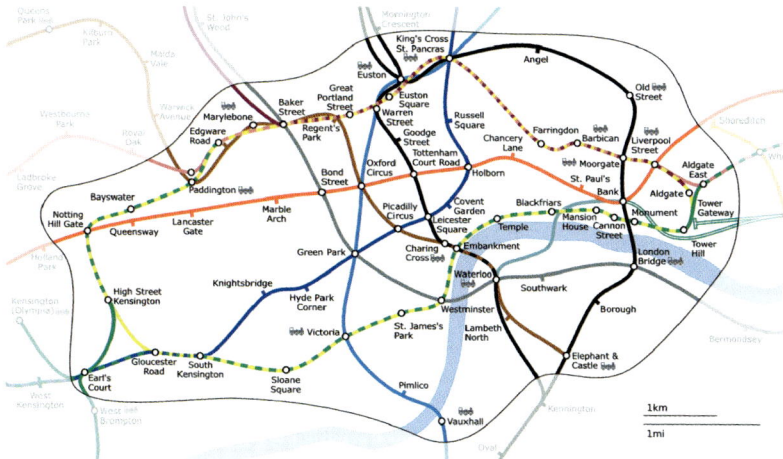

Fig. 2.6. A London subway map. According to the works of Brancolini, Buttazzo, Paolini, Pratelli, Santambrogio, Stepanov and Solimini described in Section 2.5.2, the urban transportation model involves two or more competing transportation means. In the basic model, the subway, or any fast network, is modeled as a connected set with finite length and very low cost for the users. Users walk to this network and then use it. Thus the problem is a mixed problem involving a Monge-Kantorovich individual transport to and from the fast network ($\alpha = 1$). The fast transportation means has instead a Steiner energy (the length, $\alpha = 0$).

the high speed network. The cost of the high speed network is proportional to its length and almost free to the users. The low speed network is more costly and is modeled by a Monge-Kantorovich transportation: Thus the users are led to take the shortest path to and from the transportation network. The whole problem can be viewed as a Monge-Kantorovich problem with the high-speed network as free boundary. A notion of Wasserstein distance between measures, relative to a given transportation network, is therefore introduced. The cost to be minimized is the overall transportation time for all users. The authors prove existence of high speed networks ensuring a minimal cost transportation for all users.

The transportation network is modeled as a connected closed set Σ. The users can either walk or join and use Σ. Thus, the cost for going from x to y is $d_\Sigma(x,y) := d(x,y) \wedge (dist(x, \Sigma) + dist(y, \Sigma))$, i.e. the minimum between the Euclidean (walking) distance $d(x,y)$ and the sum of distances from x and y to the network. Notice that the distance d_Σ describes how the Euclidean distance is twisted by the network. Calling the population density μ^+ and the density of workplaces μ^-, the cost of this transportation network is given by the Monge-Kantorovich distance between μ^+ and μ^-, for the cost $d_\Sigma(x,y)$. The authors consider optimal transportation networks, i.e. transportation networks with a minimal cost among all transportation networks with length less than a prescribed length L. An existence theory is presented in [17] and in [24] while [81] studies the qualitative topological and geometrical properties

of optima. It is worthwhile noticing that the above problem combines the
two extremal exponents for the Gilbert-Steiner energy, namely $\alpha = 1$ for the
pedestrians (Monge-Kantorovich) and $\alpha = 0$ for the transportation network
($\alpha = 1$).

In [70] (see also [80]), a wide extension of the Gilbert model discussed in
these notes was presented which also involves more than one transportation
means. Taking the notations of these authors and of Xia [94], consider a
one-dimensional real flat chain T such that $\partial T = \mu^+ - \mu^-$ and define its
$M^\alpha(T)$ cost as the integral over the current of the multiplicity to the power α.
Minimizing this energy only yields the generalized Gilbert energy used in [94].
Now there will be two transportation means. One of them is modeled as a
first current S with a usage cost $M^\beta(S)$ and a construction cost $H(M^\delta(S))$
where H is itself concave increasing and unbounded, thus typically a power
s^η, $0 < \eta < 1$. The second current T corresponds to another transportation
means with no construction cost and usage cost $M^\alpha(T)$. The constraint on
the irrigated measure reads $\partial(S + T) = \mu^+ - \mu^-$. The overall energy to
minimize with this constraint is

$$E(S, T) = aM^\alpha(T) + bM^\beta(S) + H(M^\delta(S)).$$

The authors prove the existence of a pair of optimal flat chains S and T
and the existence of a threshold θ_0 such that the multiplicity of S is above θ_0
and the multiplicity of T is below θ_0. In addition, $S + T$ has no cycles. To prove
the existence theorem, the one-dimensional currents are handled as measures
on a space of Lipschitz paths, that is, traffic plans in the terminology of the
present book.

In [18], an attempt was made to formulate transportation models as geo-
desic paths in the set of probability measures on a given open set U in
Euclidean space. First, a cost is defined on the space of static probability
measures. Such costs can lead to concentration or dispersion as well. In a setting very close to the Gilbert-Steiner energy, the measure under consideration
is an atomic measure $\sum_k m_k \delta_{x_k}$ and the cost is defined as $J(\mu) = \sum_k m_k^\alpha$.
Such a cost is finite only on discrete atomic masses. Then the cost of a path
$\mu(t)$ in the set of measures is the integral of its speed multiplied by this
cost, namely $\mathcal{J} = \int_0^1 J(\mu(t))|\mu'(t)|dt$ where $|\mu'(t)|$ is the metric derivative of
$t \mapsto \mu(t)$ in the Wasserstein space of measures $W_p(U)$. Such integral functionals permit to model the evolution of a measure which is forced to concentrate
on a finite set of points at any time between arrival and departure. Thus, such
costs can model the branching structure of transportation networks. Quite
interesting in this aspect is the fact that the same model permits to model a
diffusive propagation as well, forbidding any concentration, by taking $\alpha \geq 1$.
In all cases an existence result for optimal paths has been given by the authors under conditions linking the dimension and the exponent in the cost
functional. In the case $0 < \alpha < 1$ one can wonder whether the functional is a
Gilbert energy or not. In fact, it turns out to be different and has the draw-
back that the transportation cost is maintained for stopped masses, until the

whole motion of all masses has stopped. Thus, the model suffers from excessive coordination between moving masses. See also [76], for an overview of variational problems on probability measures for functionals involving transport costs with extra terms encouraging or discouraging concentration. This paper looks for optimality conditions, regularity properties and explicit computations in the case where Wasserstein distances and interaction energies are considered.

3 Traffic Plans

In this chapter, directly inspired from [11], we define almost all notions used throughout this book: traffic plan, its irrigated and irrigating measure, its transference plan, its parameterization, its fibers, the stopping time of each fiber and the multiplicity (or flow) at each point. After addressing some measurability issues we prove semicontinuity properties for all quantities of interest (average length of fibers, stopping time and multiplicity). All of these notions are first applied to the classical Monge-Kantorovich existence theory of optimal transports. Finally an existence theorem is given for traffic plans with prescribed irrigating and irrigated measures (the irrigation problem) and with prescribed transference plan (the who goes where problem).

It is convenient that the supports of the irrigated measures μ^+ and μ^- be bounded in $\mathbb{R}^N$. As we shall see (Lemma 5.15), optimal traffic plans from μ^+ to μ^- are contained in the convex hull of the union of the supports of μ^+ and μ^-. Thus it is reasonable to consider paths contained in some compact convex N-dimensional set $X \subseteq \mathbb{R}^N$. Denote by $|A|$ the Lebesgue measure of a measurable set $A \subset \mathbb{R}$.

Definition 3.1. *Let K be the set of 1-Lipschitz maps $\gamma : \mathbb{R}^+ \to X$. We define a distance on K by*

$$d(\gamma, \gamma') := \sup_{k \in \mathbb{N}^*} \frac{1}{k} ||\gamma - \gamma'||_{L^\infty([0,k])},$$

which corresponds to the metric d of uniform convergence on compact sets.

Let us denote by $\mathcal{B}$ the Borel σ-algebra on K.

Definition 3.2. *Let $\gamma \in K$. We define its stopping time as*

$$T(\gamma) := \inf\{t \geq 0 : \gamma \text{ constant on } [t, \infty[\}$$

and the length of $\gamma([0, T(\gamma)])$ by $L(\gamma)$. Since γ is 1-Lipschitz, one has

$$L(\gamma) \leq T(\gamma).$$

Remark 3.3. As we shall see in Lemma 3.20, the stopping time $T : K \to \bar{\mathbb{R}}^+$ is lower semi-continuous and therefore measurable.

M. Bernot et al., *Optimal Transportation Networks*. Lecture Notes in Mathematics 1955.

Lemma 3.4. *The metric space (K, d) is compact.*

Proof. The space K is complete and the precompactness is a straightforward consequence of Ascoli-Arzela's theorem. Let $\varepsilon > 0$ and fix k_0 so that $\sup_{k \geq k_0} \frac{\text{diameter}(X)}{k} < \varepsilon/2$. We now introduce the set $K_{k_0} \subset K$ corresponding to elements of K such that $T(\gamma) \leq k_0$. Every point in K is at a distance less than $\varepsilon/2$ from K_{k_0}. To see this, consider $\gamma \in K$. Then by restricting γ to the interval $[0, k_0]$ and extending it to $\mathbb{R}^+$ by the constant $\gamma(k_0)$, one obtains an element $\tilde{\gamma} \in K_{k_0}$ (which is still 1-Lipschitz) such that $d(\gamma, \tilde{\gamma}) \leq \sup_{k \geq k_0} \frac{\text{diameter}(X)}{k} < \varepsilon/2$.

The set K_{k_0} is equicontinuous and, for all $x \in \mathbb{R}^+$, $\{f(x) : f \in K_{k_0}\} \subset X$ is relatively compact. Thus the Ascoli-Arzela theorem ensures that K_{k_0} is relatively compact in $C([0, k_0], \mathbb{R}^N)$ equipped with $|| \cdot ||_\infty$. Within K_{k_0}, balls with radius r for the norm $|| \cdot ||_{L^\infty([0,k_0])}$ are included in balls with radius r for the distance d so that K_{k_0} is precompact in $C(\mathbb{R}^+, \mathbb{R}^N)$ equipped with d. This implies that K_{k_0} is precompact. Thus, we can consider an $\varepsilon/2$-net of K_{k_0} constituted with balls. Considering the family of balls of radius ε whose centers are the same as this $\varepsilon/2$-net, we obtain an ε-net of K.

Definition 3.5. *We define a traffic plan P as a positive measure on $(K, \mathcal{B})$ such that $\int_K T(\gamma) dP(\gamma) < \infty$. We denote by $\text{TP} = \text{TP}(X)$ the set of all traffic plans in X and by $\text{TP}_C = \text{TP}_C(X)$ the set of traffic plans P such that $\int_K T(\gamma) dP(\gamma) \leq C$.*

Remark 3.6. This definition is realistic: $T(\gamma)$ represents a transportation time and we don't want the average transportation time to be infinite! Observe also that $T(\gamma) < \infty$, P−almost everywhere.

Definition 3.7. *Consider the maps $\pi_0, \pi_\infty : K \to X$ and $\pi : K \to X \times X$ given by $\pi_0(\gamma) = \gamma(0)$, $\pi_\infty(\gamma) = \gamma(T(\gamma))$, and $\pi(\gamma) = (\gamma(0), \gamma(T(\gamma)))$. The positive measures on X and $X \times X$ defined by*

$$\mu^+(P) = \pi_{0\sharp}P, \quad \mu^-(P) = \pi_{\infty\sharp}P \text{ and } \pi(P) = \pi_\sharp P$$

will be called respectively the irrigating measure, the irrigated measure and the transference plan of P.

The notation above must be made explicit. By $\pi_{0\sharp}P$ we mean the measure on X obtained by transporting the measure P on K by the map $\pi_0 : K \to X$. In the same way $\pi_{\infty\sharp}P$ is the measure on X obtained by transporting P on X via $\pi_\infty : K \to X$. This means that for every pair of borelian sets $A, B \subset X$:

$$\pi_0(A) = P(\pi_0^{-1}(A)) = P(\{\gamma \in K : \gamma(0) \in A\});$$
$$\pi_\infty(B) = P(\pi_\infty^{-1}(B)) = P(\{\gamma \in K : \gamma(+\infty) \in B\});$$
$$\pi(A \times B) = P(\pi^{-1}(A \times B)) = P(\{\gamma \in K : \gamma(0) \in A \text{ and } \gamma(+\infty) \in B\}).$$

Thus if $\phi \in C(X, \mathbb{R})$ and $\varphi \in C(X \times X, \mathbb{R})$, one has

$$< \mu^+(P), \phi >= \int_X \phi(x) d(\pi_{0\sharp} P)(x) = \int_K \phi(\gamma(0)) dP(\gamma),$$

$$< \mu^-(P), \phi >= \int_X \phi(x) d(\pi_{\infty\sharp} P)(x) = \int_K \phi(\gamma(T(\gamma))) dP(\gamma),$$

$$< \pi(P), \varphi >= \int_{X \times X} \varphi(x_1, x_2) d(\pi_\sharp P)(x_1, x_2) = \int_K \varphi(\gamma(0), \gamma(T(\gamma))) dP(\gamma).$$

Definition 3.8. *We will call* pattern *a traffic plan* P *such that* $\mu^+(P) = \delta_S$ *for some source point* $S \in X$. *We denote by* $\mathrm{TP}(\nu^+, \nu^-)$ *the set of traffic plans* P *such that* $\mu^+(P) = \nu^+$ *and* $\mu^-(P) = \nu^-$.

3.1 Parameterized Traffic Plans

According to Skorokhod theorem (Theorem A.3 p. 185), we can parameterize any traffic plan P by a measurable function $\chi : \Omega = [0, |\Omega|] \to K$ such that $P = \chi_\sharp \lambda$, where λ is the Lebesgue measure on $[0, |\Omega|]$. We shall set $\chi(\omega, t) := \chi(\omega)(t)$ and consider it as a function of the variable pair (ω, t). This forces us to address some measurability issues.

Proposition 3.9. *Let* $\chi : \Omega \to K$. *The map* $\chi : \Omega \times \mathbb{R}^+ \to X$ *is measurable if and only if the map* $\omega \in [0, |\Omega|] \mapsto \chi(\omega, \cdot) \in K$ *is measurable.*

Proof. Assume first that $\chi : \Omega \times \mathbb{R}^+ \to X$ is measurable. Observe that

$$\chi^{-1}(B(\gamma, r)) = \{\omega : d(\chi(\omega), \gamma) \leq r\}$$
$$= \{\omega : \forall k, \frac{||\chi(\omega) - \gamma||_{L^\infty([0,k])}}{k} \leq r\}$$
$$= \cap_k \{\omega : ||\chi(\omega) - \gamma||_{L^\infty([0,k])} \leq kr\}$$
$$= \cap_k \cap_{t \in \mathbb{Q} \cap [0,k]} \{\omega : |\chi(\omega)(t) - \gamma(t)| \leq kr\}$$

This last expression is a countable intersection of measurable sets since the maps $\omega \mapsto \chi(\omega, t)$ are measurable for any t.

Conversely assume that $\chi : \Omega \to K$ is measurable. Since the map $\pi_t : K \to X$ defined by $\pi_t(\gamma) = \gamma(t)$ is continuous, the map $\omega \mapsto \pi_t \circ \chi(\omega, \cdot)$ is measurable, which means that $\omega \mapsto \chi(\omega, t)$ is measurable for every t. Let $B = B(x, r)$ be an open ball in X. By applying Lemma 3.10 below one deduces that the function $f(\omega, t) = ||\chi(\omega, t) - x||$ is measurable. Thus $f^{-1}([0, r)) = \chi^{-1}(B(x, r))$ is a measurable subset of $\Omega \times \mathbb{R}^+$.

Lemma 3.10 (see [36], [59]). *Let* $f : \Omega \times \mathbb{R}_+ \to \mathbb{R}$ *be a Caratheodory function, that is, a function such that* $\omega \mapsto f(\omega, t)$ *is measurable for each* $t \in \mathbb{R}^+$ *and that* $t \mapsto f(\omega, t)$ *is continuous for every* ω. *Then* f *is measurable with respect to the* $\sigma-$*algebra generated by the product of Lebesgue measurable sets of* Ω *and Borelians of* $\mathbb{R}^+$.

Proof. Let $D = \mathbb{Q} \cap \mathbb{R}^+$ and $a, b \in D$ be given. For any given $c > 0$ and $\varepsilon > 0$, let us introduce the sets $U = \{(\omega, t) \in \Omega \times \mathbb{R}_+ : f(\omega, t) > c\}$ and $V_\varepsilon(a, b) = \{\omega \in \Omega : f(\omega, s) > c + \varepsilon \ \forall s \in [a, b] \cap D\}$. We are going to show that

$$U = \bigcup_{\substack{a, b \in D \\ \varepsilon \in D}} V_\varepsilon(a, b) \times [a, b].$$

In order to prove the inclusion $\bigcup [a, b] \times V_\varepsilon(a, b) \subset U$, we notice that if $(\omega, t) \in [a, b] \times V_\varepsilon(a, b)$ then, for every $s \in [a, b] \cap D$, $f(\omega, s) > c + \varepsilon$ and, since we can assume $f(\omega, \cdot)$ to be continuous, by going to the limit, we get $f(\omega, t) \geq c + \varepsilon > c$.

Let us prove the reverse inclusion. If $f(\omega, t) > c$ then there exists $\varepsilon > 0$ such that $f(\omega, t) > c + 2\varepsilon$, so, by continuity, we can take a, b in such a way that for every $s \in [a, b]$ we have $f(\omega, s) > c + \varepsilon$ and $t \in [a, b]$. Now, since $f(\cdot, t)$ is measurable for every $t \in D$, then for every fixed a, b, ε, the set $V_\varepsilon(a, b)$ is measurable since it is a countable intersection of measurable sets. Therefore the above equality leads to the thesis.

Definition 3.11. *If* $\chi : \Omega \times \mathbb{R}^+ \to X$ *is measurable, we shall define its stopping time by*

$$T_\chi(\omega) := \inf\{t : \chi(\omega) \text{ is constant on } [t, \infty)\}.$$

For simplicity, we shall often write $T(\omega)$ *instead of* $T_\chi(\omega)$. *We also denote by* $L(\omega)$ *the length of the path* $t \mapsto \chi(\omega, t)$.

Remark 3.12. The measurability of $T_\chi : \Omega \to \mathbb{R}^+$ is an easy check. Indeed,

$$\{\omega : T_\chi(\omega) \leq t\} = \bigcap_{t', t'' \in \mathbb{Q}, \ t', t'' \geq t} \{\omega : \chi(\omega, t') = \chi(\omega, t'')\}$$

which is a countable intersection of measurable subsets of Ω.

Definition 3.13. *Let* Ω *be a measurable subset of* $\mathbb{R}$ *with finite measure. We call* parameterized traffic plan *a measurable map* $\chi : \Omega \times \mathbb{R}^+ \to X$ *such that* $t \mapsto \chi(\omega, t)$ *is 1-Lipschitz for all* $\omega \in \Omega$ *and* $\int_\Omega T_\chi(\omega) d\omega < +\infty$. *Without risk of ambiguity we shall call* fiber *both a path* $\chi(\omega, \cdot)$ *and the range in* $\mathbb{R}^N$ *of* $\chi(\omega, \cdot)$. *Denote by* $|\chi| := |\Omega|$ *the total mass transported by* χ *and by* $\mathbf{P}_\chi$ *the law of* $\omega \mapsto \chi(\omega) \in K$ *defined by* $\mathbf{P}_\chi(E) = |\chi^{-1}(E)|$ *for every Borel set* $E \subset K$. *Then* $\mathbf{P}_\chi$ *is a traffic plan.*

Remark 3.14. As a function of (ω, t), a parameterized traffic plan χ is a Carathéodory function. Conversely, as a consequence of Proposition 3.9, if the map χ, considered as a function of (ω, t) is a Carathéodory function, then we can define its associated parameterized traffic plan $\omega \mapsto \chi(\omega, t)$ and finally its associated traffic plan $\boldsymbol{P} := \chi_\sharp \lambda$.

Definition 3.15. *Two parameterized traffic plans χ and χ' are said to be equivalent if their laws are equal, $\mathbf{P}_\chi = \mathbf{P}_{\chi'}$.*

Since we can parameterize any traffic plan and since conversely every parameterized traffic plan defines a measure on K as its law, we will indifferently call traffic plan both kinds of object. When we say that a parameterized traffic plan has such and such property we mean that at least one representative χ in its equivalence class has it. All stated properties of a traffic plan will be meant to be true for any parameterization up to the addition or removal of a set of fibers with zero measure.

Irrigated Measures and Transference Plan in the Parametric Setting. Identifying ω with the fiber $\chi(\omega,.)$ the initial and final point of a fiber ω are $\pi_0(\omega) := \chi(\omega, 0)$ and $\pi_\infty(\omega) := \chi(\omega, T_\chi(\omega))$. The transference plan of a parameterized traffic plan χ is $\pi_\chi = (\pi_0, \pi_\infty)_\sharp \lambda$, the push forward of the Lebesgue measure λ on Ω by the map (π_0, π_∞). In more explicit terms,

$$\pi_\chi(E) = |\{\omega : (\chi(\omega, 0), \chi(\omega, T_\chi(\omega))) \in E| = |(\pi_0, \pi_\infty)^{-1}(E)|,$$

and in particular,

$$\pi_\chi(A \times B) := |\{\omega : \chi(\omega, 0) \in A,\ \chi(\omega, T_\chi(\omega)) \in B\}|.$$

Informally, $\pi_\chi(A \times B)$ represents the amount of mass transported from A to B through the traffic plan χ. In the same way we can compute the irrigating and irrigated measures of χ by

$$\mu^+(\chi)(A) := |\{\omega : \chi(\omega, 0) \in A\}|,$$

$$\mu^-(\chi)(A) := |\{\omega : \chi(\omega, T_\chi(\omega)) \in A\}|,$$

respectively, where A is any Borel subset of $\mathbb{R}^N$. Observe that $\mu^+(\chi)(A) = \pi_\chi(A \times X)$ and $\mu^-(\chi)(B) = \pi_\chi(X \times B)$. In short, $\mu^- = \pi_{0\sharp}\lambda$ and $\mu^+ = \pi_{\infty\sharp}\lambda$.

3.2 Stability Properties of Traffic Plans

Convergence

Definition 3.16. *Let $\mathbf{P}_n$ be a sequence of traffic plans. We shall say that $\mathbf{P}_n$ converges to a traffic plan $\mathbf{P}$ if one of the equivalent relations is satisfied:*

$$\mathbf{P}_n \rightharpoonup \mathbf{P},$$

$$\chi_n(\omega) \to \chi(\omega) \text{ in } K \text{ for almost all } \omega \in \Omega,$$

for some common parameterization of $\mathbf{P}_n$ and $\mathbf{P}$.

Remark 3.17. The above mentioned equivalence follows from Skorokhod theorem (see Theorem 11.7.2 [33]). The weak convergence of a sequence of positive measures $\boldsymbol{P}_n$ is equivalent to the pointwise convergence of well chosen parameterizations χ_n of $\boldsymbol{P}_n$ towards a corresponding parameterization of $\boldsymbol{P}$. A complete construction of this common parameterization can be obtained by using the common filtration constructed in Lemma A.5.

3.2.1 Lower Semicontinuity of Length, Stopping Time, Averaged Length and Averaged Stopping Time

Lemma 3.18. *(Lemma 1.61 pp. 27 [5]) Any lower semicontinuous function f on a metric compact space is the increasing limit of a sequence of continuous functions.*

Proof. Let us set $f_\varepsilon(x) := \inf_y\{f(y) + \frac{d(x,y)}{\varepsilon}\}$. Then $f_\varepsilon(x) \leq f(x)$ (take $y = x$) and f_ε is a non decreasing family of functions when ε decreases to 0. In addition, for all x, $f_\varepsilon(x)$ converges to $f(x)$. Indeed, when ε converges to 0, and for x_ε achieving the infimum, x_ε converges to x (otherwise $\frac{d(x,y)}{\varepsilon}$ would go to infinity). Thus,

$$\liminf_\varepsilon f_\varepsilon(x) = \liminf_\varepsilon \left(f(x_\varepsilon) + \frac{d(x_\varepsilon, x)}{\varepsilon} \right) \geq \liminf_\varepsilon f(x_\varepsilon) \geq f(x).$$

The last inequality comes from the lower semicontinuity of f.

Lemma 3.19. *Let $\boldsymbol{P}_n$ be a sequence of positive measures on a compact metric space K such that $\boldsymbol{P}_n$ weakly converges to $\boldsymbol{P}$. Let $\gamma \mapsto f(\gamma)$ be a lower semicontinuous function on K. Then,*

$$\int_K f(\gamma)d\boldsymbol{P}(\gamma) \leq \liminf \int_K f(\gamma)d\boldsymbol{P}_n(\gamma).$$

Proof. This is a straightforward application of Lemma 3.18 and the monotone convergence theorem. Indeed, because of Lemma 3.18, there exists a sequence of continuous functions f_k increasingly converging to f. We then have

$$\liminf_n \int_K f_k(\gamma)d\boldsymbol{P}_n(\gamma) \leq \liminf_n \int_K f(\gamma)d\boldsymbol{P}_n(\gamma).$$

The weak convergence of $\boldsymbol{P}_n$ to $\boldsymbol{P}$ yields

$$\int_K f_k(\gamma)d\boldsymbol{P}(\gamma) = \lim_n \int_K f_k(\gamma)d\boldsymbol{P}_n(\gamma).$$

Combining both relations yields

$$\int_K f_k(\gamma)d\boldsymbol{P}(\gamma) \leq \liminf_n \int_K f(\gamma)d\boldsymbol{P}_n(\gamma)$$

and one concludes by the monotone convergence theorem.

Lemma 3.20. *If the sequence $\gamma_n \in K$ converges to γ for the metric d, then*

$$T(\gamma) \leq \liminf T(\gamma_n) \ and \ L(\gamma) \leq \liminf L(\gamma_n).$$

Proof. For all $t \geq s > \liminf T(\gamma_n)$, there exists an increasing sequence of indices n_k going to infinity such that $T(\gamma_{n_k}) < s \leq t$. This ensures that $\gamma_{n_k}(t) = \gamma_{n_k}(s)$. Considering the limit of this equality, we obtain $\gamma(t) = \gamma(s)$. Then γ is constant on $(\liminf T(\gamma_n), +\infty)$, so that $T(\gamma) \leq \liminf T(\gamma_n)$. The lower semicontinuity of the length functional is well-known and we shall omit the details.

Lemma 3.21. *If a sequence of traffic plans $\boldsymbol{P}_n$ converges to $\boldsymbol{P}$, then*

$$\int_K T(\gamma)d\boldsymbol{P}(\gamma) \leq \liminf \int_K T(\gamma)d\boldsymbol{P}_n(\gamma)$$

and

$$\int_K L(\gamma)d\boldsymbol{P}(\gamma) \leq \liminf \int_K L(\gamma)d\boldsymbol{P}_n(\gamma).$$

Proof. Because of Lemma 3.20, the applications $\gamma \mapsto T(\gamma)$ and $\gamma \mapsto L(\gamma)$ are lower semicontinuous. The desired inequalities then directly come from Lemma 3.19.

3.2.2 Multiplicity of a Traffic Plan and its Upper Semicontinuity

Definition 3.22. *Let $\chi : \Omega \times \mathbb{R}^+ \to X$ be a traffic plan parameterizing $\boldsymbol{P}$. Define the path class of $x \in \mathbb{R}^N$ in χ as the set*

$$\Omega_x^\chi := \{\omega : x \in \chi(\omega, \mathbb{R})\},$$

and the multiplicity of χ at x by

$$|x|_\chi = |\Omega_x^\chi| = \boldsymbol{P}(\{\gamma : \exists t, \gamma(t) = x\}) = |x|_{\boldsymbol{P}}.$$

For simplicity, we shall write $\Omega_x := \Omega_x^\chi$, whenever the underlying traffic plan χ is not ambiguous. We denote by S_χ (resp. $S_{\boldsymbol{P}}$) the support of the traffic plan defined as the set of points x such that $|x|_\chi > 0$.

Remark 3.23. The multiplicity is well defined since the set $\{\gamma : \exists t, \gamma(t) = x\}$ is a Borel set of K. Indeed, $\{\gamma : \exists t, \gamma(t) = x\} = \cup_n\{\gamma : \exists t \leq n, \gamma(t) = x\}$ is a union of closed sets in K.

Proposition 3.24. *(Lemma 6.2, [59]) Let χ_n be a sequence of traffic plans converging to χ. Assume that $\int_\Omega T(\chi_n(\omega))d\omega \leq C$ for some C. Then, for almost all ω,*

$$\limsup |\chi_n(\omega, t)|_{\chi_n} \leq |\chi(\omega, t)|_\chi.$$

Proof. Since almost all fibers of χ converge uniformly, we shall only consider in the following fibers ω and ω' which converge uniformly. For a sake of simplicity let us adopt the abbreviation $\Omega_x^\chi = [x]_\chi$. With this notation, $|x|_\chi = |[x]_\chi|$. Set $\varepsilon = \frac{C}{M|\Omega|}$. By Markov's inequality,

$$|\{\omega : T(\chi_n(\omega)) > M\}| \leq \frac{C}{M|\Omega|} = \varepsilon.$$

Let us define an approximate multiplicity by

$$[\chi(\omega, t)]_\chi^\varepsilon := \{\omega' \in [\chi(\omega, t)]_\chi : T(\chi(\omega')) \leq M = \frac{\varepsilon|\Omega|}{C}\}.$$

Next, let us take an element ω' in $\cap_k \cup_{n>k} [\chi_n(\omega, t)]_{\chi_n}^\varepsilon$. This means that there exists a sequence of indices n_i which goes to infinity, and times $s_i \leq T(\chi_{n_i}(\omega)) \leq M$ such that $\chi_{n_i}(\omega', s_i) = \chi_{n_i}(\omega, t)$. Since s_i is bounded, it is possible to extract $s_i \to s$ and because of the uniform convergence of $\chi_{n_i}(\omega', \cdot)$ on $[0, M]$, we obtain $\chi(\omega', s) = \chi(\omega, t)$, hence $\omega' \in [\chi(\omega, t)]_\chi$. This shows that $\cap_k \cup_{n>k} [\chi_n(\omega, t)]_{\chi_n}^\varepsilon \subset [\chi(\omega, t)]_\chi$, so that

$$\limsup_n |[\chi_n(\omega, t)]_{\chi_n}^\varepsilon| \leq |[\chi(\omega, t)]_\chi|.$$

Thus,

$$\limsup_n |[\chi_n(\omega, t)]_{\chi_n}| - \varepsilon \leq |[\chi(\omega, t)]_\chi|.$$

Lemma 3.25. *Let χ be a parametrization of a traffic plan P. Then, the function $\phi : x \mapsto |x|_\chi$ is upper semicontinuous.*

Proof. Let us show that for each x such that $|x|_\chi < r$, there is a ball $B(x, \varepsilon)$ such that for all y in $B(x, \varepsilon)$, $|y|_\chi < r$. This will prove that $\phi^{-1}([0, r[)$ is an open set and therefore that ϕ is upper semicontinuous. Suppose that it is not the case. Then, for each ball $B_n := B(x, 1/n)$, there is a $y_n \in B_n$ such that $|y_n|_\chi \geq r$. Notice that y_n tends to x when n goes to infinity. Let us consider

$$\tilde{\Omega} := \cap_n \cup_{m \geq n} [y_m]_\chi.$$

(We adopt the same abbreviation as in Proposition 3.24, $\Omega_x^\chi = [x]_\chi$.) Then, modulo a null set, $\tilde{\Omega} \subset [x]_\chi$. Indeed, for almost every ω, $T(\chi(\omega)) < \infty$. For such an ω in $\tilde{\Omega}$, this means that for all n, there is an $m \geq n$ such that $\omega \in [y_m]_\chi$, that is, there is a t_m such that $\chi(\omega, t_m) = y_m$. Since $T(\chi(\omega)) < \infty$, the sequence $(t_m)_m$ can be supposed to be bounded. Thus, it is possible to extract a converging subsequence $t_m \to t$ such that $\chi(\omega, t) = x$, i.e., $\omega \in [x]_\chi$. Thus $|\tilde{\Omega}| \leq |x|_\chi < r$ and $|\tilde{\Omega}| = \lim_n |\cup_{m \geq n} [y_n]_\chi| \geq r$. This contradicts our initial assumption.

Corollary 3.26. *Let χ be a parametrization of a traffic plan P. The function $(\omega, t) \mapsto |\chi(\omega, t)|_\chi$ is measurable.*

Proof. This an immediate consequence of the measurabilities of $x \mapsto |x|_\chi$ (Lemma 3.25) and χ.

3.2.3 Sequential Compactness of Traffic Plans

Proposition 3.27. *If $(P_n)_n$ is a sequence of* TP_C *such that* $P_n \rightharpoonup P$*, then* $\pi_{P_n} \rightharpoonup \pi_P$*.*

Proof. If $P_n(K)$ tends to zero, we are done. Otherwise, up to the removal of the finite number of indices such that $P_n(K) = 0$, we can assume that P_n are probability measures by replacing P_n by $\frac{P_n}{P_n(K)}$. Set $\varepsilon = \frac{C}{M}$. By Markov's inequality, we have $P_n(K \setminus K_\varepsilon) \leq \frac{C}{M} = \varepsilon$ where $K_\varepsilon := \{\gamma : T(\gamma) \leq M\}$. Because of Lemma 3.21, we also have that $\int_K T(\gamma) dP(\gamma) \leq C$, and, thus, $P(K \setminus K_\varepsilon) < \varepsilon$. Let $\phi \in C(X \times X, \mathbb{R})$. Since, by definition of the distance on K, the map $\gamma \mapsto \phi(\gamma(0), \gamma(M))$ is continuous from K to $\mathbb{R}$, then, by definition of the transference plan associated with a traffic plan, we have

$$\limsup_n < \pi_{P_n}, \phi > \leq \limsup_n \left(\int_{K_\varepsilon} \phi(\gamma(0), \gamma(T(\gamma))) dP_n(\gamma) + \varepsilon ||\phi||_\infty \right)$$

$$= \limsup_n \int_{K_\varepsilon} \phi(\gamma(0), \gamma(M)) dP_n(\gamma) + \varepsilon ||\phi||_\infty$$

$$\leq \limsup_n \int_K \phi(\gamma(0), \gamma(M)) dP_n(\gamma) + 2\varepsilon ||\phi||_\infty$$

$$= \int_K \phi(\gamma(0), \gamma(M)) dP(\gamma) + 2\varepsilon ||\phi||_\infty$$

$$\leq \int_K \phi(\gamma(0), \gamma(T(\gamma))) dP(\gamma) + 4\varepsilon ||\phi||_\infty$$

$$= < \pi_P, \phi > + 4\varepsilon ||\phi||_\infty.$$

In the same way,

$$\liminf_n < \pi_{P_n}, \phi > \geq < \pi_P, \phi > - 4\varepsilon ||\phi||_\infty.$$

Since this is true for all $\varepsilon > 0$, the proposition is proved.

We can summarize the above results in a compactness theorem.

Theorem 3.28. *Given a bounded sequence of traffic plans* $(P_n)_n$ *in* TP_C *it is possible to extract a subsequence converging to some* P*. In addition,* $\mu^+(P_n) \rightharpoonup \mu^+(P)$*,* $\mu^-(P_n) \rightharpoonup \mu^-(P)$ *and* $\pi_{P_n} \rightharpoonup \pi_P$*.*

Proof. By extracting a subsequence we can assume that $P_n(K)$ converges. If it converges to zero we are done. Otherwise we can assume as in the previous proposition that P_n are probability measures on K. By the weak compactness theorem of probabilities on a compact metric space (see e.g. Theorem 1.59 page 26 [5]), there is a subsequence such that $P_n \rightharpoonup P$. Because of Lemma 3.21, $\int_K T(\gamma) dP(\gamma) \leq \liminf \int_K T(\gamma) dP_n(\gamma) \leq C$ so that P is a traffic plan. The end of the theorem follows from Proposition 3.27.

Corollary 3.29. *Let π be a measure on $X \times X$. There exists a traffic plan $\mathbf{P}$ such that $\pi_{\mathbf{P}} = \pi$.*

Proof : Let us first prove this property in the case of finite atomic measures π. Let $(a_i)_{i=1}^k$ and $(b_j)_{j=1}^l$ the elements of the support of the two marginals of π. Let us denote by $\pi_{i,j}$ the values $\pi(\{a_i\} \times \{b_j\})$. We now define $\gamma_{i,j} \in K$, the segment joining a_i to b_j, i.e. $\gamma_{i,j}(0) = a_i$ and, for $t \in]0, |a_i - b_j|]$,

$$\gamma_{i,j}(t) := \frac{t}{|a_i - b_j|} b_j + \frac{1-t}{|a_i - b_j|} a_i.$$

The traffic plan $\mathbf{P} := \sum_{i,j} \pi_{i,j} \delta_{\gamma_{i,j}}$ is such that $\pi_{\mathbf{P}} = \pi$ by construction.

Let us now consider a general transference plan π and a sequence of atomic measures π_n such that $\pi_n \rightharpoonup \pi$. The first part of the proof tells that there are traffic plans $\mathbf{P}_n$ such that $\pi_{\mathbf{P}_n} = \pi_n$. By theorem 3.27, we can extract a converging subsequence from $(\mathbf{P}_n)_n$ such that $\mathbf{P}_n$ converges to $\mathbf{P}$ with $\pi_{\mathbf{P}_n} \rightharpoonup \pi_{\mathbf{P}}$. Thus, the traffic plan $\mathbf{P}$ is such that $\pi_{\mathbf{P}} = \pi$. $\square$

3.3 Application to the Monge-Kantorovich Problem

For a sake of completeness let us show that the above formalism is adapted to solve the Monge-Kantorovich problem.

Definition 3.30. *We call cost of a traffic plan a functional*

$$I(\mathbf{P}) = \int_K c(\gamma(0), \gamma(T(\gamma))) d\mathbf{P}(\gamma),$$

where $c(x, y)$ is a bounded non-negative lower semicontinuous function which informally represents the cost for transporting a unit of mass from x to y.

The Monge-Kantorovich problem is the following: given two positive measures ν^+ and ν^- with equal mass, find a transference plan π that minimizes $\int_{X \times X} c(x, y) d\pi$. By definition of $\pi_{\mathbf{P}}$, we notice that $I(\mathbf{P}) = \int_{X \times X} c(x, y) d\pi_{\mathbf{P}}$. In addition, any transference plan can be obtained as the transference plan of some traffic plan as stated by Corollary 3.29. Thus, the problem of minimizing $I(\mathbf{P})$ under prescribed marginal measures $\mu^+(\mathbf{P}) = \nu^+$ and $\mu^-(\mathbf{P}) = \nu^-$ is equivalent to the Monge-Kantorovich transport problem. Without loss of generality we can assume that the masses of $\nu^{\pm}$ are 1 and consider only traffic plans such that $\mathbf{P}(K) = 1$.

Proposition 3.31. *If $(\mathbf{P}_n)_n$ and $\mathbf{P}$ are traffic plans with mass one in TP_C such that $\mathbf{P}_n \rightharpoonup \mathbf{P}$, then*

$$I(\mathbf{P}) \leq \liminf I(\mathbf{P}_n).$$

Proof. The application $\gamma \mapsto c(\gamma(0), \gamma(M))$ is lower semicontinuous because of the lower semicontinuity of c. Then Lemma 3.19 asserts that

$$\liminf_n \int_K c(\gamma(0), \gamma(M)) d\boldsymbol{P}_n(\gamma) \geq \int_K c(\gamma(0), \gamma(M)) d\boldsymbol{P}(\gamma). \qquad (3.1)$$

Set $\varepsilon = C/M$. By Markov's inequality, $\boldsymbol{P}_n(K \setminus K_\varepsilon) \leq \frac{C}{M} = \varepsilon$ where

$$K_\varepsilon := \{\gamma : T(\gamma) \leq M\}.$$

For such an M, we have

$$\int_K c(\gamma(0), \gamma(M)) d\boldsymbol{P}_n(\gamma) \leq I(\boldsymbol{P}_n) + \varepsilon ||c||_\infty$$

and

$$\int_K c(\gamma(0), \gamma(M)) d\boldsymbol{P}(\gamma) \geq I(\boldsymbol{P}) - \varepsilon ||c||_\infty,$$

so that by (3.1),

$$\liminf_n I(\boldsymbol{P}_n) + \varepsilon ||c||_\infty \geq I(\boldsymbol{P}) - \varepsilon ||c||_\infty.$$

Proposition 3.32. *The problem of minimizing $I(\boldsymbol{P})$, with $\boldsymbol{P} \in \mathrm{TP}_C(\nu^+, \nu^-)$ admits a solution for C large enough.*

Proof. We first need to prove that there is a feasible solution. An easy way is to use Skorokhod's theorem A.3, p. 185 to ν^+ and ν^-. We get parameterizations $\varphi^\pm(\omega)$ on $\Omega = [0, 1]$ for $\nu^\pm$, satisfying $\nu^\pm = \varphi_\sharp^\pm \lambda$. Since $c(x, y)$ is bounded, a feasible traffic plan is defined by $\chi(\omega, s) = s\varphi^+(0) + (1 - s)\varphi^-(0)$. Let $\boldsymbol{P}_n$ be a minimizing sequence of traffic plans, necessarily with mass one. Because of Theorem 3.27, there exists a subsequence such that $\boldsymbol{P}_n \rightharpoonup \boldsymbol{P}$ and $\pi_{\boldsymbol{P}_n} \rightharpoonup \pi_{\boldsymbol{P}}$. In particular, we have $\mu_n^+ \rightharpoonup \mu^+$ and $\mu_n^- \rightharpoonup \mu^-$. Since $\mu_n^+ = \nu^+$ and $\mu_n^- = \nu^-$ for all n, $\boldsymbol{P}$ is a traffic plan satisfying the constraints and such that $I(\boldsymbol{P}) \leq \liminf_n I(\boldsymbol{P}_n)$. Since $\boldsymbol{P}_n$ is a minimizing sequence, $\boldsymbol{P}$ is a minimizer of I under the constraints of irrigating and irrigated measures.

3.4 Energy of a Traffic Plan and Existence of a Minimizer

Our aim is to extend the Gilbert energy, only known for finite graphs, to traffic plans. The proof that the extension is consistent will be delayed until Proposition 4.8. We use the convention that $0^{\alpha-1} = \infty$ with $\alpha \in [0, 1]$.

Definition 3.33. *Let $\alpha \in [0, 1]$. We call (Gilbert) energy of a traffic plan $\boldsymbol{P}$ parameterized by χ the functional*

$$\mathcal{E}^\alpha(\boldsymbol{P}) = \int_\Omega \int_{\mathbb{R}^+} |\chi(\omega, t)|_\chi^{\alpha-1} |\dot{\chi}(\omega, t)| dt d\omega. \qquad (3.2)$$

Remark 3.34. With the above mentioned convention, let us point out that the traffic plan defined on $[0,1] \times \mathbb{R}^+$ by $\chi(\omega, t) = (\omega, t)$ for $t \leq 1$ and $\chi(\omega, t) = (\omega, 1)$ for $t \geq 1$ has infinite energy. Clearly the energy rules out a set of paths with positive measure and zero-multiplicity.

Remark 3.35. The application $(\omega, t) \mapsto |\chi(\omega, t)|_\chi$ was shown to be measurable in Corollary 3.26. The application $(\omega, t) \mapsto |\dot{\chi}(\omega, t)|$ is to be interpreted as $\liminf_{h \to 0} \frac{|\chi(\omega, t+h) - \chi(\omega, t)|}{h}$, so that measurability is ensured.

Remark 3.36. The energy of a traffic plan can also be written

$$\mathcal{E}^\alpha(\boldsymbol{P}) = \int_K \int_{\mathbb{R}^+} |\gamma(t)|_{\boldsymbol{P}}^{\alpha-1} |\dot{\gamma}(t)| dt d\boldsymbol{P}(\gamma).$$

Thus it is independent from the choice of the parameterization.

Remark 3.37. Let $\boldsymbol{P}$ be a traffic plan with finite $\mathcal{E}^\alpha$ energy. Then almost all fibers ω have a positive multiplicity almost everywhere on their domain $[0, T(\omega)]$. In particular, if fibers are parameterized by the length, fibers have positive multiplicity $\mathcal{H}^1$-almost everywhere. This motivates the following definition.

Definition 3.38. *Let $\boldsymbol{P}$ be a traffic plan parameterized by χ. The fiber ω is said to be an essential fiber if $\chi(\omega, t)$ has positive multiplicity almost everywhere on its domain $[0, T(\omega)]$.*

In the following, without loss of generality, a parameterized traffic plan with finite energy will be supposed to be made only of essential fibers.

The optimization problems we are interested in are the *irrigation problem* and the *who goes where problem*, i.e. the problem of minimizing $\mathcal{E}^\alpha(\chi)$ in $\mathrm{TP}(\mu^+, \mu^-)$ and in $\mathrm{TP}(\pi)$ respectively. The irrigation problem is obviously less constrained than the who goes where problem. Without loss of generality, we assume in the following that all measures have mass one.

Lemma 3.39. *Let $\boldsymbol{P}$ be a traffic plan with mass one. Then,*

$$\mathcal{E}^\alpha(\boldsymbol{P}) \geq \int_K L(\gamma) d\boldsymbol{P}(\gamma).$$

Proof. As the multiplicity at a point x is always less than 1, we have $|x|_{\boldsymbol{P}}^{\alpha-1} \geq 1$ and then

$$\mathcal{E}^\alpha(\boldsymbol{P}) \geq \int_K \int_{\mathbb{R}^+} |\dot{\gamma}(t)| dt d\boldsymbol{P}(\gamma) = \int_K L(\gamma) d\boldsymbol{P}(\gamma).$$

Proposition 3.40. *If $(\boldsymbol{P}_n)_n$ is a sequence in TP_C of traffic plans with mass one such that $\boldsymbol{P}_n \rightharpoonup \boldsymbol{P}$, then*

$$\mathcal{E}^\alpha(\boldsymbol{P}) \leq \liminf_n \mathcal{E}^\alpha(\boldsymbol{P}_n).$$

Proof. Let χ_n, χ be parameterizations of $\boldsymbol{P}_n$ and $\boldsymbol{P}$, respectively, such that $\chi_n(\omega) \to \chi(\omega)$ converges in (K, d) for almost every $\omega \in [0, 1]$. Because of the upper semicontinuity of multiplicity which was proved in Proposition 3.24 and the lower semicontinuity of $L(\gamma)$, we have

$$
\begin{aligned}
\liminf_n \mathcal{E}^\alpha(\boldsymbol{P}_n) = \liminf_n &\int_\Omega \int_0^{L(\chi_n(\omega))} |\chi_n(\omega, t)|_{\chi_n}^{\alpha-1} dt d\omega \\
&\geq \int_\Omega \int_0^{L(\chi(\omega))} |\chi(\omega, t)|_\chi^{\alpha-1} dt d\omega \\
&\geq \int_\Omega \int_0^{L(\chi(\omega))} |\chi(\omega, t)|_\chi^{\alpha-1} |\dot\chi(\omega, t)| dt d\omega \\
&= \mathcal{E}^\alpha(\chi) = \mathcal{E}^\alpha(\boldsymbol{P}).
\end{aligned}
$$

Proposition 3.41. *Consider two positive measures ν^+ and ν^- on X with equal mass. Assume that there exists at least one traffic plan connecting ν^+ to ν^- with finite energy. Then the problem of minimizing $\mathcal{E}^\alpha(\boldsymbol{P})$ in $\mathrm{TP}(\nu^+, \nu^-)$ admits a solution. In the same way, the problem of minimizing $\mathcal{E}^\alpha(\boldsymbol{P})$ in $\mathrm{TP}(\pi)$ admits a solution if there is at least one traffic plan $\boldsymbol{P}$ such that $\pi_{\boldsymbol{P}} = \pi$.*

Proof. By assumption there exists a traffic plan χ such that $\mathcal{E}^\alpha(\chi) < \infty$. Then there is a minimizing sequence of traffic plans $\chi_n \in \mathrm{TP}(\nu^+, \nu^-)$ which we may assume to be parameterized by arc length (see Lemma 4.2). Then for n large enough we have

$$
\int_\Omega T(\chi_n(\omega)) \, d\omega \leq \mathcal{E}^\alpha(\chi_n) \leq C := \mathcal{E}^\alpha(\chi) + 1.
$$

By Theorem 3.28 we may assume that the minimizing sequence converges to a traffic plan $\chi \in \mathrm{TP}(\nu^+, \nu^-)$. By Proposition 3.40 we deduce that χ minimizes $\mathcal{E}^\alpha(\boldsymbol{P})$ in $\mathrm{TP}(\nu^+, \nu^-)$. The last assertion is proved in a similar way.

Definition 3.42. *A traffic plan χ is said to be* optimal for the irrigation prob-lem, *respectively* optimal for the who goes where problem *if it is of minimal cost in* $\mathrm{TP}(\mu^+(\chi), \mu^-(\chi))$, *respectively in* $\mathrm{TP}(\pi_\chi)$.

Of course an optimal traffic plan χ for the irrigation problem is optimal for the who goes where problem with its own π_χ prescribed. Statements about *optimal traffic plans* mean that the properties are true no matter whether the traffic plan is optimal for the irrigation or for the who goes where problem.

4 The Structure of Optimal Traffic Plans

This chapter gives a more geometric form to traffic plans and their energy which was first proved in [59], [11]. Section 4.1 shows that all arcs of a traffic plan can be parameterized by length without changing the energy. Section 4.3.1 shows that traffic plans with finite energy are countably rectifiable. This first regularity result is used in Section 4.3 to obtain the consistency of the traffic plan energy $\mathcal{E}^\alpha$ with the original Gilbert energy. One has

$$\mathcal{E}^\alpha(\boldsymbol{P}) = E^\alpha(\boldsymbol{P}) := \int_{\mathbb{R}^N} |x|_\chi^\alpha \, d\mathcal{H}^1(x) \tag{4.1}$$

for simple paths traffic plans. Section 4.2 gives a method to actually remove the loops from any traffic plan, therefore decreasing the multiplicity $|x|_\chi$. As apparent in (4.1) the Gilbert energy decreases with the multiplicity. Thus we get a handy and intuitive Gilbert energy formula for optimal traffic plans.

Could we have directly defined the energy of a traffic plan by the right hand of (4.1)? A little thought shows that this definition would make little sense. We would have to specify first that all curves of the traffic plan belong to a countably rectifiable set, not a requirement easy to deal with. Otherwise the Gilbert energy could be zero on a spread set of curves. Next, a direct proof of the lower semicontinuity of the Gilbert energy does not seem to be easily attainable. Indeed, $x \mapsto |x|_\chi$ is not l.s.c. with respect to traffic plan convergence. Formula (4.1) will prove of great utility to the regularity analysis of the next chapters. The rectifiability of traffic plans with bounded energy is proven here by elementary techniques. A more sophisticated view, adopted by Xia [94] is the following: By Theorem 9.11 [91], when the M^α mass of a current is finite with $\alpha < 1$, then the current is rectifiable. This directly yields the rectifiability of the optimal transportation network, when $\alpha < 1$.

4.1 Speed Normalization

The following Lemma will be used twice in this Section. The proof will be given in the Appendix of this chapter.

Lemma 4.1. *Let $\chi : [0,1] \to K$ be a parameterization of the traffic plan $\boldsymbol{P}$. Let $S : [0,1] \to \mathbb{R}^+$ be such that $S(\omega, \cdot) \in \mathrm{Lip}_1(\mathbb{R}^+, \mathbb{R}^+)$ is a nondecreasing function and let $\tau : [0,1] \times [0,\infty) \to \mathbb{R}^+$ be the left inverse map*

$$\tau(\omega, s) := \inf\{t \in [0,\infty) : \ S(\omega, t) = s\}.$$

M. Bernot et al., *Optimal Transportation Networks*. Lecture Notes in Mathematics 1955.

Assume that τ is measurable. Then $\tilde{\chi}(\omega, t) = \chi(\omega, \tau(\omega, t))$ is also measurable.

Lemma 4.2. *Let $\chi : [0, 1] \to K$ be a parameterization of the traffic plan $\mathbf{P}$. Let*

$$S(\omega, t) = \int_0^t |\dot{\chi}(\omega, r)| \, dr,$$

and let

$$T(\omega, s) = \inf\{t \in [0, \infty) : S(\omega, t) = s\}.$$

Let $\tilde{\chi}(\omega, s) = \chi(\omega, T(\omega, s))$. Then $\tilde{\chi}$ is Lebesgue measurable and each fiber is parameterized by length. We call $\tilde{\chi}$ the arc-length normalization of χ. If $\chi = \tilde{\chi}$, we shall say that χ is normalized. One has $\mathcal{E}^\alpha(\tilde{\chi}) = \mathcal{E}^\alpha(\chi)$.

Proof. By Lemma 4.1, the measurability of $\tilde{\chi}$ is a consequence of the measurability of T. To prove that $T^{-1}((-\infty, \lambda])$ is measurable for any $\lambda \in \mathbb{R}$, let $\{t_m\}_m$ be a dense sequence in $[0, \infty)$. Using that T is non decreasing and lower semicontinuous in s we may write

$$T^{-1}((-\infty, \lambda]) = \bigcap_{n=1}^\infty \bigcup_{m=1}^\infty \{\omega \in [0, 1] : T(\omega, t_m) \le \lambda\} \times [0, t_m + \frac{1}{n}].$$

Since $\{\omega \in [0, 1] : T(\omega, t_m) \le \lambda\} = \{\omega \in [0, 1] : S(\omega, \lambda) \ge t_m\}$ is measurable, we deduce that $T^{-1}((-\infty, \lambda])$ is measurable.

By definition of $T(\omega, s)$, we have

$$\int_0^{T(\omega, s)} |\dot{\chi}(\omega, r)| \, dr = s. \tag{4.2}$$

Observe that for a.e. $\omega \in [0, 1]$ we have

$$|\tilde{\chi}(\omega, s + h) - \tilde{\chi}(\omega, s)| = |\chi(\omega, T(\omega, s + h)) - \chi(\omega, T(\omega, s))|$$
$$\le \int_{T(\omega, s)}^{T(\omega, s+h)} |\dot{\chi}(\omega, r)| \, dr = h,$$

for any $s, s + h \in [0, \infty)$, hence $\tilde{\chi}(\omega, \cdot)$ is a 1-Lipschitz function. Since $\tilde{\chi}(\omega, \cdot)$ is the composition of a Lipschitz function $\chi(\omega, \cdot)$ with a nondecreasing function $T(\omega, \cdot)$, we have that

$$\dot{\tilde{\chi}}(\omega, s) = \dot{\chi}(\omega, T(\omega, s)) \frac{\partial T}{\partial s}(\omega, s) \qquad \text{a.e. in } s.$$

Now, for each fixed ω, $\int_0^{T(\omega, s)} |\dot{\chi}(\omega, r)| \, dr$ is the composition of a Lipschitz function $t \mapsto \int_0^t |\dot{\chi}(\omega, r)| \, dr$ and a nondecreasing function $s \mapsto T(\omega, s)$. The composition is a function of bounded variation, hence is differentiable almost everywhere. Moreover, because of (4.2), its derivative has no singular part. Thus, we have that

$$|\dot{\chi}(\omega, T(\omega, s))| \frac{\partial T}{\partial s}(\omega, s) = 1 \text{ a.e.,} \qquad (4.3)$$

which means $|\dot{\tilde{\chi}}(\omega, s)| = 1$. The fact that $\mathcal{E}^{\alpha}(\tilde{\chi}) = \mathcal{E}^{\alpha}(\chi)$ is a straightforward consequence of (4.3) and the fact that the multiplicity is unchanged by arc-length normalization, namely $|\tilde{\chi}(\omega, s)|_{\tilde{\chi}} = |\chi(\omega, T(\omega, s))|_{\chi}$.

Definition 4.3. *We say that $\tilde{P}$ is a normalization of a traffic plan P if it is the law of the arc-length renormalization $\tilde{\chi}(\omega)$ of some parameterization $\chi(\omega)$ of P. Observe that $\mathcal{E}^{\alpha}(\tilde{P}) = \mathcal{E}^{\alpha}(P)$.*

Remark 4.4. Due to the fact that $\{\gamma \in K : |\dot{\gamma}| = 1\}$ is not closed under the distance d, it is not true that $P_n \rightharpoonup P$ implies $\tilde{P}_n \rightharpoonup \tilde{P}$.

4.2 Loop-Free Traffic Plans

Definition 4.5. *A traffic plan P is said to be loop-free if there is a parameterization χ of P so that for almost all $\omega \in [0, 1]$, the element $\chi(\omega)$ of K is injective on $[0, T(\omega)]$.*

Proposition 4.6. *Let P be a traffic plan such that $\mathcal{E}^{\alpha}(P) < \infty$. There exists a loop-free traffic plan $\tilde{P}$ such that $S_{\tilde{P}} \subset S_P$, $|x|_{\tilde{P}} \leq |x|_P$, for all x, and $\pi_{\tilde{P}} = \pi_P$. This traffic plan is obtained by removing iteratively the largest loops from all fibers.*

Proof. Since the support and the transference plans are invariant under normalization of a traffic plan P, we can suppose P to be normalized. Let χ be a parameterization of P. Because of Lemma 3.39, $L(\chi(\omega)) < \infty$ for almost all $\omega \in \Omega$. For these ω, we reparameterize the path $\chi(\omega)$ after having removed its loops. To do so, we introduce the set

$$X_{\omega} = \{x \in \chi(\omega, \mathbb{R}^+) : \text{Card}(\chi(\omega, \cdot)^{-1}(x) \cap [0, L(\chi(\omega))]) > 1\},$$

which is empty if and only if $\chi(\omega)$ is injective.

Step 1: Existence of a maximal set of injectivity. We shall call a set of injectivity a set

$$A_{\omega} = \bigcup_{x \in X_{\omega}} [t_x^-, t_x^+[$$

such that $\chi(\omega)$ is injective on $[0, L(\chi(\omega))] \setminus A_{\omega}$, where t_x^- and t_x^+ are elements of $\chi(\omega, \cdot)^{-1}(x)$.

Let us use an iterative process to construct such a set. Consider first the set $T_{\omega}^0 = [0, L(\chi(\omega))]$. If $\chi(\omega)$ is injective on T_{ω}^0, then the empty set is a set of injectivity. Otherwise, we consider one of the largest intervals $[t_1^-, t_1^+[$ where t_1^- and t_1^+ are in $T_{\omega}^0 \cap \chi(\omega, \cdot)^{-1}(x)$ with x in X_{ω}. Such an interval exists since $[0, L(\chi(\omega))]$ is bounded. We then set $T_\omega^1 = T_\omega^0 \setminus [t_1^-, t_1^+[$. Continuing this process iteratively, we obtain a decreasing sequence of sets

$$T_\omega^n = T_\omega^{n-1} \setminus [t_n^-, t_n^+[,$$

where $t_n^-, t_n^+ \in T_\omega^{n-1} \cap \chi(\omega, \cdot)^{-1}(x)$ and $x \in X_\omega$. The process stops whenever $\cup_{k=1}^n [t_k^-, t_k^+[$ is a set of injectivity. If the process never ends, the set $\cup_{k=1}^\infty [t_k^-, t_k^+[$ is a set of injectivity. Indeed, let us assume that $s_1, s_2 \in [0, L(\omega)] \setminus \cup_k [t_k^-, t_k^+[$ are such that $\chi(\omega, s_1) = \chi(\omega, s_2)$. Then, by construction,

$$\infty > L(\chi(\omega)) \geq \sum_n |t_n^+ - t_n^-| \geq \sum_n |s_1 - s_2|,$$

thus $s_1 = s_2$. We shall denote by T_ω the set $[0, L(\omega)] \setminus \cup_k [t_k^-, t_k^+[$.

Step 2: Definition of the reparameterization. The set T_ω is a set of time parameters describing an injective subpath of $\chi(\omega)$. Let us consider the non-decreasing continuous function

$$S_\omega(u) = \int_0^u \mathbb{1}_{T_\omega}(s) ds$$

and let us define $\tau_\omega(t) := \inf\{u \in [0, \infty) : S_\omega(u) = t\}$. Then, $\tau_\omega(t)$ is such that $|T_\omega \cap [0, \tau_\omega(t)]| = t$ for any $t \in [0, |T_\omega|]$. We observe that the map $\tau_\omega(t)$ is measurable as a function of (ω, t). The proof of this fact will be given in Lemma 4.11 in the appendix of this chapter. We reparameterize the paths $\chi(\omega, s)$ by $\tilde{\chi}(\omega, t) := \chi(\omega, \tau_\omega(t))$. Observe that, by Lemma 4.1, $\tilde{\chi}$ is measurable. We can then define $\tilde{P} := \tilde{\chi}_\sharp \lambda$.

Step 3: The traffic plan $\tilde{P}$ is loop-free. Indeed, if there is an ω such that $\tilde{\chi}(\omega)$ is not injective, there are t_1 and t_2 such that $y = \tilde{\chi}(\omega, t_1) = \tilde{\chi}(\omega, t_2)$ with $t_1 \neq t_2$. Then, since τ_ω is increasing, $\tau_\omega(t_1) \neq \tau_\omega(t_2)$. Thus $\text{Card}\chi_\omega^{-1}(y) > 1$ so by definition of A_ω one of these two elements has to be in A_ω. But this is not possible since the image of τ_ω is disjoint from A_ω. Thus, $\tilde{\chi}$ is loop-free. By definition of $\tilde{\chi}$, $\pi_{\tilde{P}} = \pi_P$. Since $\tilde{\chi}$ is a restriction of χ, the multiplicity of $\tilde{\chi}$ is less than the multiplicity of χ at every point and in particular $S_{\tilde{P}} \subset S_P$.

4.3 The Generalized Gilbert Energy

Definition 4.7. *Let P be a traffic plan and χ a parameterization of P. For each $\omega \in \Omega$, we define*

$$\mathcal{D}^\chi(\omega) = \{x \in \mathbb{R}^N : x \text{ is a double point of } \chi(\omega)\}.$$

We say that χ has simple paths if $\mathcal{H}^1(\mathcal{D}^\chi(\omega)) = 0$ for almost every $\omega \in \Omega$.

Assume that for a given $\omega \in \Omega$, $\chi(\omega)$ is parameterized by arc-length. Let

$$\mathcal{D}_\chi(\omega) = \{t \in [0, \infty) : \exists s < t, \ \chi(\omega, t) = \chi(\omega, s)\}.$$

Observe that $\mathcal{H}^1(\mathcal{D}^\chi(\omega)) = 0$ if and only if $|\mathcal{D}_\chi(\omega)| = 0$. Thus, if χ is normalized, χ has simple paths if and only if $|\mathcal{D}_\chi(\omega)| = 0$ for almost every $\omega \in \Omega$.

The following representation of the energy shows that the traffic plan energy is a generalization of the Gilbert energy.

Proposition 4.8. *Let* $\alpha \in [0, 1)$ *and* χ *be a parameterization of a traffic plan* **P** *with finite energy. Then*

$$\mathcal{E}^\alpha(\boldsymbol{P}) = \int_\Omega \int_0^\infty |\chi(\omega, t)|_\chi^{\alpha-1} |\dot{\chi}(\omega, t)| \, dt \, d\omega \geq \int_{\mathbb{R}^N} |x|_\chi^\alpha \, d\mathcal{H}^1(x). \qquad (4.4)$$

If we assume in addition that χ *has simple paths we have*

$$\mathcal{E}^\alpha(\boldsymbol{P}) = \int_\Omega \int_0^\infty |\chi(\omega, t)|_\chi^{\alpha-1} |\dot{\chi}(\omega, t)| \, dt \, d\omega = \int_{\mathbb{R}^N} |x|_\chi^\alpha \, d\mathcal{H}^1(x). \qquad (4.5)$$

Proof. The simple argument below was pointed out to us by Filippo Santambrogio. If for almost every ω the path $t \to \chi(\omega, t)$ has almost no double point, the push-forward by $\chi(\omega, \cdot)$ of $|\dot{\chi}(\omega, t)|dt$ on $[0, T(\omega)]$ is nothing but the $\mathcal{H}^1$ measure restricted to the range of $\chi(\omega, \cdot)$, which we denote by $R(\omega)$. Thus, using Fubini-Tonelli Theorem,

$$\mathcal{E}^\alpha(\boldsymbol{P}) = \int_\Omega \int_0^{T(\omega)} |\chi(\omega, t)|_\chi^{\alpha-1} |\dot{\chi}(\omega, t)| dt d\omega = \int_\Omega \int_{R(\omega)} |x|_\chi^{\alpha-1} d\mathcal{H}^1(x) d\omega$$

$$= \int_\Omega \int_{\mathbb{R}^N} |x|_\chi^{\alpha-1} \mathbb{1}_{x \in R(\omega)} d\mathcal{H}^1(x) d\omega = \int_{\mathbb{R}^N} |x|_\chi^{\alpha-1} \int_\Omega \mathbb{1}_{x \in R(\omega)} d\omega d\mathcal{H}^1(x)$$

$$= \int_{\mathbb{R}^N} |x|_\chi^{\alpha-1} |x|_\chi d\mathcal{H}^1(x) = \int_{\mathbb{R}^N} |x|^\alpha d\mathcal{H}^1(x).$$

This yields (4.5). Observe that we are using the convention that $0^{\alpha-1} = \infty$ and it is implicitly understood that the support of the integral consists of the points $x \in \mathbb{R}^N$ where $|x|_\chi > 0$. In case χ is not simple-path, the second equality in the above series becomes an inequality, and we get (4.4).

From now on we denote the intrinsic formulation of the energy by

$$E^\alpha(\boldsymbol{P}) = \int_{\mathbb{R}^N} |x|_{\boldsymbol{P}}^\alpha \, d\mathcal{H}^1(x).$$

Proposition 4.9. *The minimum of* $\mathcal{E}^\alpha$ *on the set of traffic plans, if it exists, is attained at a loop-free traffic plan. Moreover* $\inf \mathcal{E}^\alpha = \inf E^\alpha$, *where both infima are taken indifferently over the set of all traffic plans or over the set of loop-free traffic plans.*

Proof. We prove that if P is a traffic plan and $\tilde{P}$ its associated loop-free traffic plan constructed in Proposition 4.6, we have $\mathcal{E}^\alpha(\tilde{P}) \leq \mathcal{E}^\alpha(P)$. To prove it, we observe that when eliminating loops, the multiplicity decreases, hence $E^\alpha(P) \geq E^\alpha(\tilde{P})$. Now, by Proposition 4.8, we have

$$\mathcal{E}^\alpha(P) \geq E^\alpha(P) \geq E^\alpha(\tilde{P}) = \mathcal{E}^\alpha(\tilde{P}).$$

Our assertions are a simple consequence of Proposition 4.6 and this inequality.

4.3.1 Rectifiability of Traffic Plans with Finite Energy

Theorem 4.10. *Let P be a traffic plan with finite energy and χ a parameterization of P. Assume that almost all fibers have positive length. Then S_χ is countably rectifiable. More precisely, there exists a sequence $(\omega_j)_j$ such that*

$$|x|_\chi = 0 \quad \mathcal{H}^1 - \text{a.e., for} \quad x \in \mathbb{R} \setminus \cup_{j=1}^\infty \text{Im } \chi(\omega_j). \tag{4.6}$$

Proof. Proposition 4.8 implies that the set S_χ of points x such that $|x|_\chi > 0$ has σ−finite $\mathcal{H}^1$-measure. Thus, by a classical result of Besicovich (see [39]), S_χ will be countably rectifiable if its $\mathcal{H}^1$-density superior to 1 for almost every point in S_χ. We now prove that it is the case. From the definition 3.33 of the energy we know that there is $\Omega' \subset \Omega$ with $|\Omega'| = 1$ such that for every $\omega \in \Omega'$ the multiplicity of $\chi(\omega, t)$ is positive at almost every point of $[0, T_\chi(\omega)]$. Consider the restriction χ' of χ to $\cup_{\omega \in \Omega'}\{\omega\} \times (0, T_\chi(\omega))$. Notice that the support of χ' is entirely contained in open essential fibers. Obviously $\mathcal{E}^\alpha(\chi) = \mathcal{E}^\alpha(\chi')$ and $\int_\Omega T_\chi(\omega) = \int_{\Omega'} T_\chi(\omega)$, which implies that $\int_{\mathbb{R}^N} |x|_\chi d\mathcal{H}^1 = \int_{\mathbb{R}^N} |x|_{\chi'} d\mathcal{H}^1$ and therefore that $|x|_\chi = |x|_{\chi'}$ $\mathcal{H}^1$-almost everywhere. It follows that $\mathcal{H}^1$-almost every point in S_χ is contained in some (open) fiber ω of χ'. Thus this fiber crosses the whole ball $B(x, r)$ for r small enough, which implies that the $\mathcal{H}^1$-density of S_χ in $B(x, r)$ tends to 1.

4.4 Appendix: Measurability Lemmas

Let us give the proof of Lemma 4.1.

Proof. The map $\tilde{\chi}$ is the composition of the maps $(I, \tau) : [0, 1] \times [0, \infty) \to [0, 1] \times [0, \infty)$ and $\chi : [0, 1] \times [0, \infty) \to \mathbb{R}^N$. The measurability of $\tilde{\chi}$ will be a consequence of the measurability of (I, τ) and χ, and the fact that $(I, \tau)^{-1}(N)$ is a null set in $[0, 1] \times [0, \infty)$ for any null set N in $[0, 1] \times [0, \infty)$. The measurability of (I, τ) follows from the measurability of τ.

Now, let N be a null set in $[0, 1] \times [0, \infty)$ and let B be a Borel set containing N (of total measure less than ε). Observe that $F(\omega, s) := \mathbb{1}_B(\omega, \tau(\omega, s))$ is a measurable map. For a.e. fixed value of each $\omega \in [0, 1]$, we have

$$\int_0^\infty F(\omega, s)ds = \int_0^\infty \mathbb{1}_B(\omega, t)S_t(\omega, t)dt \leq \int_0^\infty \mathbb{1}_B(\omega, t)dt,$$

the last inequality being true since $S_t(\omega, t) \leq 1$. Integrating with respect to $\omega \in [0, 1]$, and observing that both F and $\mathbb{1}_B$ are measurable in $[0, 1] \times [0, \infty)$, we have

$$|(I, \tau)^{-1}(B)| = \int_0^1 \int_0^\infty \mathbb{1}_B(\omega, \tau(\omega, s)) ds d\omega \leq \int_0^1 \int_0^\infty \mathbb{1}_B(\omega, t) dt d\omega \leq \varepsilon.$$

We deduce that $(I, \tau)^{-1}(N)$ is a null set.

Lemma 4.11. *The map $\tau_\omega(t)$ defined in Lemma 4.6 is measurable as a function of (ω, t).*

Proof. Let $\{t_m\}$ be a dense sequence in $[0, \infty)$. Following the proof of Lemma 4.2, since $\tau_\omega(t)$ is non-decreasing, lower semicontinuous, and

$$\{\omega \in [0, 1] : \tau_\omega(t_m) \leq \lambda\} = \{\omega \in [0, 1] : S_\omega(\lambda) \geq t_m\}$$

it suffices to prove that the sets $\{\omega \in [0, 1] : S_\omega(\lambda) \geq t_m\}$ are measurable for any $\lambda \geq 0$. For that, it is sufficient to prove that the sets

$$S = \{\omega \in [0, 1] : S_\omega(\lambda) \leq t_m\} = \{\omega \in [0, 1] : |T_\omega \cap [0, \lambda]| \leq t_m\}$$
$$= \{\omega \in [0, 1] : |T_\omega^c \cap [0, \lambda]| \geq \lambda - t_m\}$$

are measurable for any $\lambda \geq 0$. Let

$$T_{\omega, p} = [0, L(\omega)] \setminus \cup_{\{k : t_k^+ - t_k^- \geq \frac{1}{p}\}} [t_k^-, t_k^+[$$

and observe that $\cap_p T_{\omega, p} = T_\omega$ Let us prove that for any $p \geq 1$, the set

$$S_p := \{\omega \in [0, 1] : |T_{\omega, p}^c \cap [0, \lambda]| \geq \lambda - t_m\}$$

is measurable. Recall that, since $\chi : [0, 1] \to K$ is measurable, for each $j \in \mathbb{N}$, there is a compact set $B_j \subseteq [0, 1]$ such that $\chi : B_j \to K$ is continuous. Let us prove that for any $j \in \mathbb{N}$ the set

$$S_{p,j} := \{\omega \in [0, 1] : |T_{\omega, p}^c \cap [0, \lambda]| \geq \lambda - t_m\} \cap B_j$$

is closed, hence, a Borel set. Let $\omega_i \in S_{p,j}$, $\omega_i \to \omega$. Then, for each of the curves $\chi(\omega_i)$, the sum of the lengths of the loops of length $\geq \frac{1}{p}$ is $\geq \lambda - t_m$. Letting $i \to \infty$, we deduce that the sum of the lengths of the loops of $\chi(\omega)$ of length $\geq \frac{1}{p}$ is also $\geq \lambda - t_m$. In other words, $\omega \in S_{p,j}$. Since $S_p = \cup_j S_{p,j} \cup N$ where N is a null set, we deduce that S_p is a measurable set. Now, since $\cup_p T_{\omega, p}^c = T_\omega^c$, we have that

$$\{\omega : |T_\omega^c \cap [0, \lambda]| \geq \lambda - t_m\} = \{\omega : \sup_p |T_{\omega, p}^c \cap [0, \lambda]| \geq \lambda - t_m\}$$

$$= \bigcap_j \bigcup_k \{\omega : |T_{\omega, k}^c \cap [0, \lambda]| \geq \lambda - t_m - \frac{1}{j}\}.$$

Hence S is measurable. We conclude that $\tau_\omega(t)$ is measurable as a function of (ω, t).

5 Operations on Traffic Plans

This chapter is devoted to a series of elementary operations permitting to combine traffic plans into new ones. New traffic plans can be built by restriction, concatenation, union, and hierarchical concatenation. These technical tools were introduced in [13]. The hierarchical concatenation gives a standard way to explicitly construct infinite irrigation trees, or patterns. Incidentally three elementary assumptions will be justified by their lack of incidence on the search of optimal traffic plans: X can be taken convex, the mass of χ can always be taken equal to one, and we can get rid of zero-length fibers.

5.1 Elementary Operations

5.1.1 Restriction, Domain of a Traffic Plan

Definition 5.1. *Let χ be a parameterized traffic plan on $\Omega \times \mathbb{R}^+$. If $\Omega' \subset \Omega$ we call restriction of χ to $\Omega' \times \mathbb{R}^+$, the traffic plan $\chi_{|\Omega' \times \mathbb{R}^+}$ also noted $\chi_{\Omega'}$. More generally, let $\Omega' \subset \Omega$ and $D \subset \Omega' \times \mathbb{R}^+$ a subset of the form $D = \cup_{\omega \in \Omega'} \{\omega\} \times [s(\omega), t(\omega)]$. Define the restriction $\chi_{|D}$ of χ to D as a traffic plan by $\chi_D(\omega, t) = \chi(\omega, t + s(\omega))$ if $0 \le t \le t(\omega) - s(\omega)$ and $\chi_D(\omega, t) = \chi(\omega, t(\omega))$ if $t \ge t(\omega) - s(\omega)$.*

Lemma 5.2. *(i) Let $D = \cup_{\omega \in \Omega'} \{\omega\} \times [s(\omega), t(\omega)]$. Then*

$$E^\alpha(\chi_{|D}) \le E^\alpha(\chi). \tag{5.1}$$

(ii) If Ω is a disjoint union of $\Omega_1, \ldots, \Omega_l$, then

$$E^\alpha(\chi) \le \sum_{i=1}^{l} E^\alpha(\chi_{|\Omega_i}).$$

Proof. (i) Follows immediately from the definition of E^α because $|x|_{\chi_D} \le |x|_\chi$.

(ii) Observe that for every point x one has $\Omega_x^\chi = \cup_i \Omega_x^{\chi_{|\Omega_i}}$ and the union is disjoint. Thus by the subadditivity of $s \mapsto s^\alpha$,

$$E^\alpha(\chi) = \int_{\mathbb{R}^N} |x|_\chi^\alpha d\mathcal{H}^1 \le \int_{\mathbb{R}^N} \sum_{i=1}^{l} |x|_{\chi_{|\Omega_i}}^\alpha \, d\mathcal{H}^1 = \sum_{i=1}^{l} E^\alpha(\chi_{|\Omega_i}).$$

M. Bernot et al., *Optimal Transportation Networks*. Lecture Notes in Mathematics 1955.

47

For many results it is convenient to have all fibers of a traffic plan parameterized on $[0, T(\omega)]$. However, we shall also call traffic plan a map χ defined on some set $D = \cup_{\omega \in \Omega} \{\omega\} \times [s(\omega), t(\omega)]$, as arises naturally when we perform a restriction of a traffic plan. We call *domain of a traffic plan* χ the set

$$\mathcal{D}(\chi) = \cup_{\omega \in \Omega} \{\omega\} \times [S(\omega), T(\omega)] \tag{5.2}$$

where $T(\omega)$ is the stopping time of the fiber and $S(\omega)$ its starting time (usually 0).

5.1.2 Sum of Traffic Plans (or Union of their Parameterizations)

Definition 5.3. *Let P_i be a sequence of traffic plans such that $\sum_i P(K) < \infty$ and $(\chi_i)_{i \in I}$ parameterizations of P_i defined on $\Omega_i \times \mathbb{R}^+$ with $|\Omega_i| = P_i(K)$. We call sum of the traffic plans P_i the traffic plan $P = \sum_i P_i$. This traffic plan can be given a parameterization from the χ_i's as follows. Let $\Omega = [0, \sum_{i \in I} |\Omega_i|]$. Take a partition of Ω into sets $\tilde{\Omega}_i$ such that $|\tilde{\Omega}_i| = |\Omega_i|$ for all $i \in I$. Let $\phi_i : \tilde{\Omega}_i \to \Omega_i$ be measure preserving applications. Consider the "union" of the parameterized traffic plans χ_i denoted by $\cup_{i \in I} \chi_i$ and defined on $\Omega \times \mathbb{R}^+$ by $(\cup_{i \in I} \chi_i)(\omega, t) = \chi_j(\phi_j(\omega), t)$, for $\omega \in \tilde{\Omega}_j$. Obviously $\cup_{i \in I} \chi_i$ is a parameterization of $\sum_i P_{\chi_i}$.*

5.1.3 Mass Normalization

Definition 5.4. *Let χ be a traffic plan. Up to a measure preserving application, we can suppose that χ is defined on $\Omega = [0, |\chi|]$. Define the renormalization of χ as the traffic plan $\tilde{\chi} : [0, 1] \times \mathbb{R}^+ \to \mathbb{R}^N$ such that $\tilde{\chi}(\omega, t) = \chi(|\chi|\omega, t)$. In that way, $|\tilde{\chi}| = 1$.*

Unless otherwise specified, we always assume that the considered traffic plans are parameterized by length. This leads to no loss of generality since one can normalize every traffic plan.

5.2 Concatenation

5.2.1 Concatenation of Two Traffic Plans

Lemma 5.5. *Let $\chi \in \mathrm{TP}(\mu^+, \mu^-)$ and $\xi \in \mathrm{TP}(\nu^+, \nu^-)$ such that $\mu^- = \nu^+$. There is a traffic plan $\tilde{\chi} \in \mathrm{TP}(\mu^+, \nu^-)$ such that each fiber of $\tilde{\chi}$ is a concatenation of a fiber of χ with a fiber of ξ. In addition, $\mathcal{E}^\alpha(\tilde{\chi}) \leq \mathcal{E}^\alpha(\chi) + \mathcal{E}^\alpha(\xi)$ and $E^\alpha(\tilde{\chi}) \leq E^\alpha(\chi) + E^\alpha(\xi)$.*

Proof. Let us denote $f(\omega) := \chi(\omega, T(\omega))$ and $g(\omega) := \xi(\omega, 0)$. By definition, $\mu^- = \nu^+$ means that $f_\sharp \lambda = g_\sharp \lambda$. Thus, there is a measure preserving application $\psi : \Omega \to \Omega$ such that $f(\omega) = g(\psi(\omega))$ for almost all ω. The concatenation of the fiber $\chi(\omega)$ with the fiber $\xi(\psi(\omega))$ is a Lipschitz path and we can define the concatenated traffic plan $\tilde{\chi}$ by

$$\tilde{\chi}(\omega, t) = \begin{cases} \chi(\omega, t) & \text{if } t \leq T_\chi(\omega) \\ \xi(\psi(\omega), t - T_\chi(\omega)) & \text{if } t > T_\chi(\omega), \end{cases}$$

where $T_\chi(\omega)$ is the stopping time of the fiber ω. The traffic plan $\tilde{\chi}$ satisfies $|x|_{\tilde{\chi}} \geq |x|_\chi$ and $|x|_{\tilde{\chi}} \geq |x|_\xi$. Thus, by definition of $\mathcal{E}^\alpha$,

$$\mathcal{E}^\alpha(\tilde{\chi}) = \int_\omega \int_t |\tilde{\chi}(\omega, t)|_{\tilde{\chi}}^{\alpha-1} |\dot{\tilde{\chi}}(\omega, t)| \, d\omega dt$$

$$\leq \int_\omega \int_{t < T_\chi(\omega)} |\tilde{\chi}(\omega, t)|_\chi^{\alpha-1} |\dot{\tilde{\chi}}(\omega, t)| \, d\omega dt$$

$$+ \int_\omega \int_{t > T_\chi(\omega)} |\tilde{\chi}(\omega, t)|_\xi^{\alpha-1} |\dot{\tilde{\chi}}(\omega, t)| \, d\omega dt = \mathcal{E}^\alpha(\chi) + \mathcal{E}^\alpha(\xi).$$

The last assertion follows from inequality $|x|_{\tilde{\chi}} \leq |x|_\chi + |x|_\xi$ and the definition of E^α.

5.2.2 Hierarchical Concatenation (Construction of Infinite Irrigating Trees or *Patterns*)

Since notation here is necessarily a bit cumbersome, we refer to Figure 5.1.

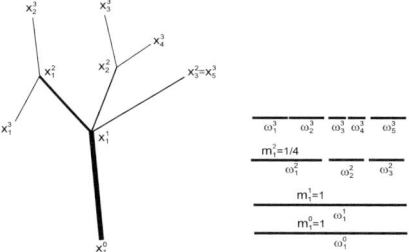

Fig. 5.1. This figure illustrates the notion of hierarchy of collectors (Definition 5.6). A hierarchy of collectors is made of sets P_i that contain points x_j^i, where $1 \leq j \leq k_i$, and surjective maps γ_i from P_i to P_{i-1}. This is a way to describe a tree structure on points x_j^i. In this particular figure, we have $P_0 = \{x_1^0\}$, $P_2 = \{x_1^2, x_2^2, x_3^2\}$. The map γ_3 describes to which point in P_2 a point of P_3 is connected. For instance, $\gamma_3(x_3^3) = x_2^2$.

Definition 5.6. *We shall say that*

$$(P_i)_{i \in \mathbb{N}} = ((x_j^i)_{i \in \{1,\ldots,k_i\}}, \gamma_i, m_i = (m_j^i)_{j \in \{1,\ldots k_i\}})_{i \in \mathbb{N}}$$

is a hierarchy of collectors *if*

- P_0 *is a single point* $S \in \mathbb{R}^N$ *(which is the "head" of the hierarchy and will be the source in the applications), and* $\forall i \in \mathbb{N}$, $i \geq 1$, P_i *is a finite subset of* $\mathbb{R}^N$ *made of* k_i *point collectors* x_j^i, $1 \leq j \leq k_i$;
- $\forall i \in \mathbb{N}, i \geq 1$, γ_i *is a surjective map from* P_i *onto* P_{i-1} *which associates with every collector* x_j^i *with rank* i *the rank* $i-1$ *collector of* $\gamma_i(x_j^i)$ *it is connected to.*
- $m_i(x_j^i) = m_j^i$ *is the mass transported from* $\gamma_i(x_j^i) \in P_{i-1}$ *to* $x_j^i \in P_i$ *and* $\sum_j m_j^i = 1$.

We assume that all collectors P_i are disjoint. Thus $m_i(x)$ is unambiguous because x belongs to at most one P_i.

Definition 5.7. *Given a hierarchy of collectors* (P_i, γ_i, m_i), *consider the transition traffic plans* Q_i *defined as the union of all elementary traffic plans following the segments* $[\gamma_i(x_j^i), x_j^i]$ *at constant speed and with flows* m_j^i. *Thus* Q_i *irrigates* $\mu_i := \sum_{j=1}^{k_i} m_j^i \delta_{x_j^i}$ *from from* μ_{i-1}. *We associate with the hierarchy of collectors the patterns* $P_{m,n}$, $n > m$, *obtained by concatenating the traffic plans* Q_i, $i = m+1, \ldots, n$. *Thus* $P_{m,n}$ *irrigates* μ_n *from* μ_m *and* $P_n := P_{0,n}$ *irrigates* μ_n *from the source* S.

Lemma 5.8. *Let* $(P_i, \gamma_i, m_i)_{i \in \mathbb{N}}$ *be a hierarchy of collectors and* $P_{m,n}$ *its associated patterns. Then*

$$E^\alpha(P_{m,n}) \leq \mathcal{E}^\alpha(P_{m,n}) \leq \sum_{i=m+1}^{n} \sum_{x \in P_i} m_i(x)^\alpha |x - \gamma_i(x)|. \qquad (5.3)$$

Proof. The cost of the elementary pattern going from $\gamma_i(x)$ to x with flow $m_i(x)$ is $m_i(x)^\alpha |x - \gamma_i(x)|$. Thus, by the estimate in Lemma 5.5,

$$\mathcal{E}^\alpha(Q_i) \leq \sum_{x \in P_i} m_i(x)^\alpha |x - \gamma_i(x)| .$$

Iterating this estimate yields (5.3).

Corollary 5.9. *Let* $(P_i, \gamma_i, m_i)_{i \in \mathbb{N}}$ *be a hierarchy of collectors and* $P_{m,n}$ *its associated patterns. Then*

$$\mathcal{E}^\alpha(P_{m,n}) \leq \sum_{i=m+1}^{n} k_i^{1-\alpha} l_i ,$$

where for all $i \in \{1, \ldots, n\}$

$$l_i = \max_{x \in P_i} |x - \gamma_i(x)| .$$

Proof. By Hölder inequality and the assumption $\sum_{x \in P_i} m_i(x) = 1$,

$$\sum_{x \in P_i} m_i(x)^\alpha \le \left(\sum_{x \in P_i} m_i(x) \right)^\alpha \left(\sum_{x \in P_i} 1 \right)^{1-\alpha} = k_i^{1-\alpha} \ .$$

Then the announced estimate follows from (5.3).

5.3 A Priori Properties on Minimizers

5.3.1 An Assumption on μ^+, μ^- and π Avoiding Fibers with Zero Length

If $\chi : \Omega \times \mathbb{R}^+ \to X$ is a traffic plan in $\mathrm{TP}(\mu^+, \mu^-)$, then we may eliminate all fibers with zero length. Indeed, if $\Omega_0 = \{\omega : \chi(\omega, t) = \chi(\omega, 0) \ \forall t \ge 0\}$, then $\chi' := \chi_{|(\Omega \setminus \Omega_0) \times \mathbb{R}^+}$ is a traffic plan in $\mathrm{TP}(\tilde{\mu}^+, \tilde{\mu}^-)$, where $\tilde{\mu}^+ = \mu^+ - \mu_0^+$ and $\tilde{\mu}^- = \mu^- - \mu_0^-$ and $\mu_0^+ = \mu_0^-$ are the irrigating and irrigated measures of $\chi_{|\Omega_0 \times \mathbb{R}^+}$. Observe that the transference plan associated to $\chi_{|\Omega_0 \times \mathbb{R}^+}$ is the restriction to the diagonal set $\{(x, x) : x \in X\}$ of the transference plan associated to χ. Observe that $\mathcal{E}^\alpha(\chi_{|(\Omega \setminus \Omega_0) \times \mathbb{R}^+}) = \mathcal{E}^\alpha(\chi)$ and, by Lemma 5.2, $E^\alpha(\chi_{|(\Omega \setminus \Omega_0) \times \mathbb{R}^+}) \le E^\alpha(\chi)$.

Lemma 5.10. *Let* $\chi : \Omega \times \mathbb{R}^+ \to X$ *be a traffic plan in* $\mathrm{TP}(\mu^+, \mu^-)$ *with finite energy,* $\alpha \in [0, 1)$. *Then there exists a traffic plan* $\tilde{\chi} \in \mathrm{TP}(\mu^+ - \mu^+ \wedge \mu^-, \mu^- - \mu^+ \wedge \mu^-)$ *such that* $E^\alpha(\tilde{\chi}) \le E^\alpha(\chi)$.

Proof. By the observations previous to this lemma, we may assume that χ has no fiber of zero length. Let $\nu := \mu^+ \wedge \mu^-$. We may assume that $\nu \ne 0$, the result being obviously true if $\nu = 0$. On the other hand we may assume that $\mu^+ - \nu \ne 0$, since otherwise $\mu^+ = \mu^-$ and the traffic plan with constant fibers connecting μ^- to μ^+ is optimal. In this case $\tilde{\chi} \in \mathrm{TP}(0, 0)$. Let S^+, S^-, S^0 be disjoint Borel sets such that $\mu^+ - \nu$ is supported in S^+, $\mu^- - \nu$ is supported in S^-, and ν is supported in S^0. By the previous observations, we may assume that S^+, S^-, S^0 have all positive measure. By neglecting a set of fibers of null measure, we may assume that all fibers $\omega \in \Omega$ start in $S^+ \cup S_0$ and arrive at $S^- \cup S^0$. Let $\alpha \in \{+, 0\}$, $\beta \in \{-, 0\}$, and let

$$\Omega_{\alpha, \beta} := \{\omega \in \Omega : \chi(\omega, 0) \in S^\alpha, \ \chi(\omega, T_\chi(\omega)) \in S^\beta\}.$$

Notice that the fibers such that $\chi(\omega, 0) = \chi(\omega, T_\chi(\omega))$ can only be in $\Omega_{0,0}$. Thus, using Proposition 4.6 (see also the proof of Proposition 4.9), by eliminating the loops of χ we get a loop-free traffic plan χ' in $TP(\mu^+ - \nu + \nu^0, \mu^- - \nu + \nu^0)$ where $\nu^0 << \nu$ such that $E^\alpha(\chi') \le E^\alpha(\chi)$. Thus, without loss of generality, we may assume that χ is a loop-free traffic plan. This implies that if A is a set of fibers of positive measure in $\Omega_{0,0}$, the irrigating and irrigated measures are positive submeasures of ν of equal mass $|A|$. Thus there cannot be infinite chains of concatenations of fibers in $\Omega_{0,0}$. Thus, if $\omega \in \Omega_{+,0}$, it can be concatenated with another fiber in $\Omega_{0,-} \cup \Omega_{0,0}$. If it is concatenated with a fiber in $\Omega_{0,0}$ after a finite number of concatenations $k \geq 0$, we arrive to $\Omega_{0,-}$. This permits to define a new traffic plan $\tilde{\chi}$ in $\mathrm{TP}(\mu^+ - \nu, \mu^- - \nu)$. Notice that this traffic plan can be indexed by fibers in $\Omega_{+,-}$. By eliminating again the loops, we may assume that $\tilde{\chi}$ is loop-free. Notice that $|x|_{\tilde{\chi}} \leq |x|_{\chi}$ for any x in the support of $\tilde{\chi}$. This implies that $E^\alpha(\tilde{\chi}) \leq E^\alpha(\chi)$, and the proof concludes.

Proposition 5.11. *Let $\alpha \in [0,1)$ and μ^+, μ^- be two positive measures of equal mass that can be connected by a traffic plan with finite energy. Let $\nu := \mu^+ \wedge \mu^-$. Then there is an optimal traffic plan in $\mathrm{TP}(\mu^+, \mu^-)$ which is the sum of an optimal traffic plan in $\mathrm{TP}(\mu^+ - \nu, \mu^- - \nu)$ and a traffic plan with constant fibers in $\mathrm{TP}(\nu, \nu)$.*

Proof. Let $\tilde{\chi}$ be optimal traffic plan in $\mathrm{TP}(\mu^+ - \nu, \mu^- - \nu)$ and let χ_0 be a traffic plan with constant fibers in $\mathrm{TP}(\nu, \nu)$, which is also optimal. Then we define $\chi \in \mathrm{TP}(\mu^+, \mu^-)$ as the sum of $\tilde{\chi}$ and χ_0. Using Lemma 5.10, it follows that χ is an optimal traffic plan in $\mathrm{TP}(\mu^+, \mu^-)$.

Remark 5.12. Notice that not all optimal traffic plans in $\mathrm{TP}(\mu^+, \mu^-)$ have the structure described in Proposition 5.11. Indeed, if μ^+, μ^- are the Lebesgue measures concentrated in the intervals $[0,2]$, $[1,3]$, respectively, then the two following traffic plans χ_1 and χ_2 are optimal. We define $\chi_1(\omega, t) = \min (\omega + t, \omega + 2)$, if $\omega \in [0,1]$, $t \in [0, +\infty)$, and $\chi_1(\omega, t) = \omega$ for any $\omega \in (1, 2]$, $t \in [0, \infty)$. Then χ_1 is optimal. The proof is essentially the same as the proof of Lemma 13.2. Let $\chi_2(\omega, t) = \min(\omega + t, \omega + 1)$ for any $\omega \in [0,2]$, $t \in [0, \infty)$. A direct computation of the energy shows that $E^\alpha(\chi_1) = E^\alpha(\chi_2) = 1 + \frac{2}{\alpha+1}$ for any $\alpha \in [0,1]$. Thus, χ_2 is also optimal.

Thus, by Proposition 5.11, there is no deep loss of generality if we assume that μ^+ and μ^- are orthogonal. In the next Proposition we notice that when $\mu^+ \perp \mu^-$, then almost all fibers have positive length.

Proposition 5.13. *Let $\mu^+ \perp \mu^-$ be two positive measures on X with equal mass. Then for every traffic plan irrigating μ^- from μ^+ almost all fibers have positive length.*

Proof. Let O be the set of fibers with null length and $\nu^+ := \pi_{\infty|O\sharp}\lambda$, $\nu^- := \pi_{0|O\sharp}\lambda$ where $\pi_0(\omega) = \chi(\omega, 0)$ and $\pi_\infty(\omega) = \chi(\omega, T(\omega))$. Then $\pi_0 = \pi_\infty$ on O implies $\nu^+ = \nu^-$. Considering that $\nu^+ << \mu^+$, $\nu^- << \mu^-$ and taking into account the assumption $\mu^+ \perp \mu^-$ yields $|\nu^+| = |\nu^-| = |O| = 0$.

Let us now consider the traffic plan optimization problem with prescribed transference plan π. Here we shall always assume without loss of generality that $\pi(\{(x,x) : x \in X\}) = 0$. Otherwise one can consider the set of self-irrigated points, namely the Borel set $O \subset X$ such that $\pi(\{(x,x) : x \in O\})$ is maximal. If χ is an optimal traffic plan irrigating π, the restriction $\chi_{|O}$ is made of zero length fibers. We can remove this spurious set and replace π by $\tilde{\pi}$ defined by $\tilde{\pi}(A \times B) := \pi((A \setminus 0) \times (B \setminus 0))$. Conversely, by the very same arguments as in Proposition 5.13, we have:

Proposition 5.14. *Let π be a transference plan such that $\pi(\{(x,x) : x \in X\}) = 0$. Then for every traffic plan irrigating π, almost all fibers have positive length.*

Proof. Let $O \subset \Omega$ the set of fibers with zero length. Then,

$$E = (\pi_0(O), \pi_\infty(O)) \subset \{(x,x) : x \in X\},$$

and therefore

$$
\begin{aligned}
|O| &= |(\pi_0, \pi_\infty)^{-1}(E)| \\
&\leq |(\pi_0, \pi_\infty)^{-1}(\{(x,x) : x \in X\})| \\
&= \pi(\{(x,x) : x \in X\}) = 0.
\end{aligned}
$$

5.3.2 A Convex Hull Property

We denote by $conv(E)$ the convex hull of E.

Lemma 5.15. *An optimal traffic plan χ satisfies $S_\chi \subset conv(\text{supp}(\mu^-(\chi)) \cup \text{supp}(\mu^+(\chi)))$. More precisely almost all fibers of the traffic plan stay in this convex hull.*

Proof. Let $\mathbf{C} := \overline{conv(\text{supp}(\mu^-(\chi)) \cup \text{supp}(\mu^+(\chi)))}$. For all $\omega \in \Omega$, define $\tilde{\chi}(\omega,t) = p_{\mathbf{C}}(\chi(\omega,t))$ where $p_{\mathbf{C}}$ denotes the projection on the convex $\mathbf{C}$. Since $\chi(\omega,0)$ and $\chi(\omega,\infty)$ belong to $\mathbf{C}$, $\tilde{\chi}$ has the same transference plan and irrigated measures as χ. Assume by contradiction that $t \mapsto \chi(\omega,t)$ goes out of $\mathbf{C}$ for ω in a set with positive measure. Since $p_{\mathbf{C}}$ is a contractive map, the function $t \mapsto p_{\mathbf{C}}(\chi(\omega,t))$ is a Lipschitz function whose derivative is almost everywhere defined and $|\frac{d}{dt}(p_{\mathbf{C}}(\chi(\omega,t)))| \leq \frac{d}{dt}|\chi(\omega,t)|$, the inequality being strict on a subset of $\mathbb{R}^+$ with positive measure if $\chi(\omega,t)$ goes out of $\mathbf{C}$ on some interval with positive measure. Obviously, $p_{\mathbf{C}}$ being an application, $|p_{\mathbf{C}}(x)|_{p_{\mathbf{C}} \circ \chi} \geq |x|_\chi$ for all $x \in \mathbb{R}^N$. Thus

$$
\begin{aligned}
E^\alpha(\tilde{\chi}) \leq \mathcal{E}^\alpha(\tilde{\chi}) &= \int_\Omega \int_{\mathbb{R}^+} |\tilde{\chi}(\omega,t)|_{\tilde{\chi}}^{\alpha-1} |\frac{d}{dt}\tilde{\chi}(\omega,t)| dt d\omega \\
&= \int_\Omega \int_{\mathbb{R}^+} |p_{\mathbf{C}}(\chi(\omega,t))|_{p_{\mathbf{C}} \circ \chi}^{\alpha-1} |\frac{d}{dt}(p_{\mathbf{C}} \circ \chi)(\omega,t)| dt d\omega \\
&< \int_\Omega \int_{\mathbb{R}^+} |\chi(\omega,t)|_\chi^{\alpha-1} |\frac{d}{dt}\chi(\omega,t)| dt d\omega = \mathcal{E}^\alpha(\chi) = E^\alpha(\chi).
\end{aligned}
$$

By Lemma 5.15 an optimal traffic plan whose irrigated measures have support in a convex set $\mathbf{C}$ is equivalent to a traffic plan contained in $\mathbf{C}$. Thus, given μ^+ and μ^- with compact support in $\mathbf{C}$, it is natural to choose $X = \mathbf{C}$. However, the choice of any larger convex set for X yields the very same optimal traffic plans.

6 Traffic Plans and Distances between Measures

In this chapter, we consider the irrigation and who goes where problems for the cost functional E^α introduced at the end of Chapter 3. We prove in Section 6.1 that for $\alpha > 1 - \frac{1}{N}$ where N is the dimension of the ambient space, the optimal cost to transport μ^+ to μ^- is finite. More precisely, if μ^+ and μ^- are two nonnegative measures on a domain X with the same total mass M and $\alpha > 1 - 1/N$, set

$$E^\alpha(\mu^+, \mu^-) := \min_{\chi \in \mathrm{TP}(\mu^+, \mu^-)} E^\alpha(\chi). \tag{6.1}$$

Then $E^\alpha(\mu^+, \mu^-)$ can be bounded by

$$E^\alpha(\mu^+, \mu^-) \le C_{\alpha, N} M^\alpha \mathrm{diam}(X).$$

The proof of this property, first proven in [94], follows from the explicit construction of a dyadic tree connecting any probability measure on X to a Dirac mass. If α is under this threshold it may happen that the infimum is in fact $+\infty$.

Section 6.3 compares E^α with the so called Wasserstein distance associated with the Monge-Kantorovich model. The sharp quantitative estimate that is obtained takes the form

$$W_1(\mu^+, \mu^-) \le E^\alpha(\mu^+, \mu^-) \le c W_1(\mu^+, \mu^-)^\beta$$

for some $\beta > 0$. The question of the existence of such an equality was raised by Cédric Villani and its proof given in [63]. This inequality gives a quantitative proof of the fact that E^α and W_1 induce the same topology on the set $\mathcal{P}(X)$ of probability measures on X. This topology is the weak convergence topology.

Because the topology induced by E^α induces the weak topology for $\alpha > 1 - \frac{1}{N}$, we have $E^\alpha(\nu_n, \nu) \to 0$ when ν_n is a sequence of probability measures on the compact $X \subset \mathbb{R}^N$ weakly converging to ν. As a consequence the limit of a converging sequence of optimal traffic plans for E^α is still optimal. This settles the stability of optima with respect to μ^+ and μ^-.

Lemma 6.1. *Let us denote W_1 the Wasserstein distance of order 1 and let μ^+ and μ^- be two probability measures. We have $W_1(\mu^+, \mu^-) \le E^\alpha(\mu^+, \mu^-)$ for all $\alpha \in [0, 1]$.*

Proof. Indeed,

$$E^\alpha(\mu^+, \mu^-) := \inf_{\chi \in \mathrm{TP}(\mu^+, \mu^-)} \int_\Omega \int_t |\chi(\omega, t)|_\chi^{\alpha - 1} |\dot\chi(\omega, t)| d\omega dt,$$

where the infimum is taken over all parameterizations transporting μ^+ to μ^-. In particular,

$$E^1(\mu^+, \mu^-) := \inf_{\chi \in \mathrm{TP}(\mu^+, \mu^-)} \int_\Omega \int_t |\dot\chi(\omega, t)| d\omega dt$$

is precisely $W_1(\mu^+, \mu^-)$. Since $|\dot\chi(\omega, t)|_\chi^{\alpha-1} \geq 1$, we obtain

$$W_1(\mu^+, \mu^-) \leq E^\alpha(\mu^+, \mu^-).$$

Proposition 6.2. E^α *is a pseudo-distance on the space of probability measures on* X.

Proof. Because of Lemma 6.1, we have $E^\alpha(\nu_1, \nu_2) = 0$ if and only if $\nu_1 = \nu_2$. Next, the triangular inequality is easily proved as follows: let $\boldsymbol{P}_1$ and $\boldsymbol{P}_2$ be optimal traffic plans respectively from ν_1 to ν_2 and from ν_2 to ν_3. By definition of E^α, we have

$$E^\alpha(\nu_1, \nu_3) \leq E^\alpha(\boldsymbol{P}),$$

where $\boldsymbol{P}$ is the concatenation of $\boldsymbol{P}_1$ and $\boldsymbol{P}_2$ defined in Lemma 5.5. Thus

$$E^\alpha(\nu_1, \nu_3) \leq E^\alpha(\boldsymbol{P}_1) + E^\alpha(\boldsymbol{P}_2) = E^\alpha(\nu_1, \nu_2) + E^\alpha(\nu_2, \nu_3).$$

6.1 All Measures can be Irrigated for $\alpha > 1 - \frac{1}{N}$

Let C be a cube with edge length L and center c. Let ν be a probability measure on $X \subset C$. One can approximate ν by atomic measures as follows. For each i, let

$$C_j^i : j \in \mathbb{Z}^N \cap [0, 2^i)^N$$

be a partition of C into cubes of edge length $\frac{L}{2^i}$. For $j \in \mathbb{Z}^N \cap [0, 2^i)^N$ call x_j^i the center of C_j^i and let $m_j^i = \nu(C_j^i)$ be the ν-mass of the cube C_j^i.

Definition 6.3. *With the above notation we call dyadic approximation of a measure* ν *supported by a cube the atomic measure*

$$\mu_i = \mu_i(\nu) = \sum_{j \in \mathbb{Z}^N \cap [0, 2^i)^N} m_j^i \delta_{x_j^i}.$$

The following lemma is very classical.

Lemma 6.4. *The atomic measures* μ_i *weakly converge to* ν.

Lemma 6.5. *Let* ν *be a probability measure on a cube* C *of edge length* L. *Then for* $n > m$,

$$E^\alpha(\mu_m, \mu_n) \leq \mathcal{E}^\alpha(\boldsymbol{P}_{m,n}) \leq \frac{\sqrt{N}L}{2} \frac{2^{m(N(1-\alpha)-1)}}{2^{1-N(1-\alpha)} - 1}.$$

Proof. This a direct application of Corollary 5.9. The number k_i of collectors at scale 2^{-i} is equal to 2^{Ni} and the length of the segments connecting them to the collectors at scale 2^{-i+1} is equal to $l_i = L\sqrt{N}2^{-i-1}$. Thus (see Figure 6.1),

$$\mathcal{E}^\alpha(\boldsymbol{P}_{m,n}) \leq \sum_{i=m+1}^n k_i^{1-\alpha}l_i$$

$$\leq \sum_{i=m+1}^\infty \frac{L\sqrt{N}}{2}2^{i(N(1-\alpha)-1)}$$

$$= \frac{L\sqrt{N}}{2} \frac{2^{m(N(1-\alpha)-1)}}{2^{1-N(1-\alpha)} - 1}.$$

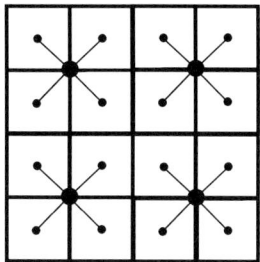

Fig. 6.1. To transport μ_i to μ_{i+1}, all the mass at the center of a cube with edge length $\frac{L}{2^{i-1}}$ is transported to the centers of its sub-cubes with edge length $\frac{L}{2^i}$.

Proposition 6.6. *Let $\alpha \in (1-\frac{1}{N}, 1]$. Let ν be a probability measure supported in a cube centered at c with edge length L. Then*

$$E^\alpha(\mu_n(\nu), \nu) \leq \frac{2^{n(N(1-\alpha)-1)}}{2^{1-N(1-\alpha)} - 1} \frac{\sqrt{N}L}{2}.$$

In particular, $E^\alpha(\mu_n(\nu), \nu) \to 0$ uniformly for all ν when $n \to \infty$

Proof. By construction, the traffic plan $\boldsymbol{P}_{m,n}$ converges to a traffic plan $\boldsymbol{P}^m$ irrigating the measure ν from μ_m. (All fibers of $\boldsymbol{P}_{m,n}$ converge uniformly to fibers whose length is less than $\sqrt{N}L$.) Thus by Lemma 6.5 and Proposition 3.40,

$$E^\alpha(\mu_m, \nu) \leq \liminf_n \mathcal{E}^\alpha(\boldsymbol{P}_{m,n}) \leq \frac{2^{n(N(1-\alpha)-1)}}{2^{1-N(1-\alpha)} - 1} \frac{\sqrt{N}L}{2}. \tag{6.2}$$

Since $\mu_0 = \delta_c$, we obtain directly from the previous proposition applied with $n = 0$ the following uniform bound on the energy required to irrigate a measure. Notice that a set with diameter L is contained in a cube with edge $2L$.

Corollary 6.7. *Let $\alpha \in (1 - \frac{1}{N}, 1]$ and ν be a probability measure on a set X with diameter L. There exists $\mathbf{P} \in \mathrm{TP}(\delta_c, \nu)$ such that*

$$E^\alpha(\mathbf{P}) \le \frac{1}{2^{1-N(1-\alpha)} - 1} \sqrt{N} L.$$

Remark 6.8. The work of Devillanova and Solimini [78] refines widely the result of Corollary 6.7 by giving precise conditions on ν to be α−irrigable (see chapter 10).

Finally, combining a transport from μ^+ to δ_c with a transport from δ_c to μ^-, it is possible to obtain any transference plan, so that the who goes where problem has a solution at finite cost in the case $\alpha > 1 - \frac{1}{N}$.

Corollary 6.9. *Let $\alpha \in (1 - \frac{1}{N}, 1]$. Let μ^+ and μ^- be probability measures on X, and π a prescribed transference plan with marginals μ^+ and μ^-. There exists $\mathbf{P} \in \mathrm{TP}(\pi)$ such that*

$$E^\alpha(\mathbf{P}) \le \frac{1}{2^{1-N(1-\alpha)} - 1} 2\sqrt{N} L.$$

Proof. Indeed, we can find a traffic plan $\mathbf{P}_1$ transporting μ^+ to δ_c and a traffic plan $\mathbf{P}_2$ transporting δ_c to μ^- such that

$$E^\alpha(\mathbf{P}_1) + E^\alpha(\mathbf{P}_2) \le \frac{2}{2^{1-N(1-\alpha)} - 1} \sqrt{N} L.$$

By concatenating $\mathbf{P}_1$ and $\mathbf{P}_2$ one obtains a traffic plan $\mathbf{P}$ with a transference plan $\pi_{\tilde{\mathbf{P}}}$ that can be any transference plan with marginal laws μ^+ and μ^-. Since $|x|_{\tilde{\mathbf{P}}} \le |x|_{\mathbf{P}_1} + |x|_{\mathbf{P}_2}$, we have

$$E^\alpha(\tilde{\mathbf{P}}) \le E^\alpha(\mathbf{P}_1) + E^\alpha(\mathbf{P}_2) \le \frac{2}{2^{1-N(1-\alpha)} - 1} \sqrt{N} L.$$

Corollary 6.10. *If the transported measure has mass M, the uniform bounds obtained in Corollaries 6.7 and 6.9 scale as M^α and we have*

$$E^\alpha(\mu^+, \mu^-) \le C_{\alpha,N} M^\alpha \mathrm{diam}(X) \tag{6.3}$$

6.2 Stability with Respect to μ^+ and μ^-

In this section we partially answer the stability question, i.e. "is the limit of a sequence of optimal traffic plans optimal?". The property of the E^α pseudo-distance in the case $\alpha \in (1 - \frac{1}{N}, 1]$ permits to answer yes (Proposition 6.12). However, in the case $\alpha \le 1 - \frac{1}{N}$ this stability is conjectural.

Lemma 6.11. *Let $\alpha \in (1 - \frac{1}{N}, 1]$. If ν_n is a sequence of probability measures on the compact $X \subset \mathbb{R}^N$ weakly converging to ν, then $E^\alpha(\nu_n, \nu) \to 0$ when $n \to \infty$.*

Proof. Let us adopt the notation of Definition 6.3 and Proposition 6.6 and let us assume that X is contained in a cube with edge length L subdivided into dyadic cubes C_j^i with edge length $2^{-i}L$. The weak convergence of ν_n to ν applied to the characteristic functions of the cubes C_j^i implies that $m_j^i(\nu_n)$ converges to $m_j^i(\nu)$ when $n \to \infty$, where $m_j^i(\nu)$ denotes the mass of ν contained in the cube C_j^i. Thus for any $\varepsilon > 0$ and for n large enough,

$$\sum_j |m_j^i(\nu_n) - m_j^i(\nu)| < \varepsilon.$$

By Proposition 6.6, $E^\alpha(\mu_i(\nu), \nu) \le \varepsilon$ and $E^\alpha(\mu_i(\nu_n), \nu_n) \le \varepsilon$ for i large enough, independently of n. We are left to evaluate $E^\alpha(\mu_i(\nu_n), \mu_i(\nu))$. Since these measures are concentrated at the centers of cubes C_j^i, this amounts to transport in the whole cube a mass less than $\sum_j |m_j^i(\nu_n) - m_j^i(\nu)| < \varepsilon$. By (6.3), we deduce that $E^\alpha(\mu_i(\nu), \mu_i(\nu_n)) \le C\varepsilon^\alpha$ for a constant C depending only on X and α. The triangular inequality for E^α yields

$$E^\alpha(\nu, \nu_n) \le E^\alpha(\nu, \mu_i(\nu)) + E^\alpha(\mu_i(\nu), \mu_i(\nu_n)) + E^\alpha(\mu_i(\nu_n), \nu_n) \le 2\varepsilon + C\varepsilon^\alpha.$$

Proposition 6.12. *Let $\alpha \in (1 - \frac{1}{N}, 1]$. If $\mathbf{P}_n$ is a sequence of optimal traffic plans for the irrigation problem and $\mathbf{P}_n$ is converging to $\mathbf{P}$, then $\mathbf{P}$ is optimal.*

Proof. Since $E^\alpha(\mathbf{P}_n) = \mathcal{E}^\alpha(\mathbf{P}_n)$ and $E^\alpha(\mathbf{P}) \le \mathcal{E}^\alpha(\mathbf{P})$, using the lower semi-continuity of $\mathcal{E}^\alpha$, we have

$$\begin{aligned}
E^\alpha(\mathbf{P}) &\le \liminf_n E^\alpha(\mathbf{P}_n) = \liminf_n E^\alpha(\mu_n^+, \mu_n^-) \\
&\le \liminf_n \left(E^\alpha(\mu_n^+, \mu^+) + E^\alpha(\mu^+, \mu^-) + E^\alpha(\mu^-, \mu_n^-) \right) \\
&\le E^\alpha(\mu^+, \mu^-) \text{ since } \mu_n^+ \to \mu^+ \text{ and } \mu_n^+ \to \mu^+.
\end{aligned}$$

Thus, $\mathbf{P}$ is optimal.

Remark 6.13. In the case $\alpha < 1 - \frac{1}{N}$, the stability of optimal traffic plans remains an open question (see Chapter 15). Of course, only the case when $\mathbf{P}_n$ is a sequence of optimal traffic plans with $E^\alpha(\mathbf{P}_n) < \infty$ is of interest. The stability in the case of the who goes where problem is also an open problem.

6.3 Comparison of Distances between Measures

Proposition 6.12 implies that the topology induced by the distance E^α on $\mathcal{P}(X)$ is exactly the weak-* topology.

Proposition 6.14. *If $\alpha \in (1 - \frac{1}{N}, 1]$, E^α is a metric of the weak-$*$ topology of probability measures $\mathcal{P}(X)$.*

Proof. Indeed, Proposition 6.12 asserts that if ν_n weakly converges to ν then $E^\alpha(\nu_n, \nu) \to 0$. Conversely, if $E^\alpha(\nu_n, \nu) \to 0$, then Lemma 6.1 asserts that $W_1(\nu_n, \nu) \to 0$, so that ν_n weakly converges to ν. $\qquad\square$

Remark 6.15. If $\alpha \leq 1 - \frac{1}{N}$, then it is no longer true that if ν_n weakly converges to ν then $E^\alpha(\nu_n, \nu) \to 0$. Indeed, let us consider $\nu_n := \frac{1}{v_n} \mathbb{1}_{B(0, \frac{1}{n})}$, where v_n is the volume of a ball with radius $\frac{1}{n}$. In that case $\nu_n \rightharpoonup \delta_0$ but, by Theorem 10.26, $E^\alpha(\nu_n, \delta_0) = \infty$ if $\alpha \leq 1 - \frac{1}{N}$.

The following proposition gives a quantitative version of Proposition 6.14. To fix ideas, we consider two probability measures μ^+ and μ^- with support in an N-dimensional cube C with edge 1, say $C = [0, 1]^N$. It is not difficult to scale the result to any bounded domain in $\mathbb{R}^N$.

Proposition 6.16. *Let $\alpha \in (1 - \frac{1}{N}, 1]$, then*

$$E^\alpha(\mu^+, \mu^-) \leq c W_1(\mu^+, \mu^-)^{N(\alpha - (1 - 1/N))},$$

where c denotes a suitable constant depending only on N and α.

Proof. Let us denote X^+ and X^- the two projections from $C \times C$ onto C, so that $X^+(x, y) = x$, $X^-(x, y) = y$.

Let π_0 be an optimal transport plan between μ^+ and μ^-, i.e. a probability measure on $C \times C$ such that $X^\pm_\# \pi_0 = \mu^\pm$ and with cost $w := W_1(\mu^+, \mu^-)$. We denote by

$$\Lambda_i = \left\{ (x, y) \in C \times C : (2^i - 1)\frac{w}{2} \leq |x - y| < (2^{i+1} - 1)\frac{w}{2} \right\}.$$

We can limit ourselves to consider those indices i which are not too large, i.e. up to $(2^i - 1)\frac{w}{2} \leq \sqrt{N}$ (where $\sqrt{N}$ is the diameter of C). Let I be the maximal index i so that this inequality is satisfied. The set $\Lambda = \cup_{i=0}^I \Lambda_i$ is a disjoint union and

$$\sum_{i=0}^I (2^i - 1)\frac{w}{2}\pi_0(\Lambda_i) \leq W_1(\mu^+, \mu^-) = w \leq \sum_{i=0}^I (2^{i+1} - 1)\frac{w}{2}\pi_0(\Lambda_i). \quad (6.4)$$

We call cube with edge e any translate of $[0, e[^N$. For each $i = 0, \cdots, I$, using a regular grid in $\mathbb{R}^N$, one can cover C with disjoint cubes $C_{i,k}$ with edge $(2^{i+1} - 1)w$. The number of the cubes in the i-th covering may be easily estimated by

$$\left(\frac{1}{(2^{i+1} - 1)w} + 1 \right)^N \leq \left(\frac{c}{(2^{i+1} - 1)w} \right)^N = K(i). \quad (6.5)$$

For each index i, C is included in the disjoint union $\subset \cup_{k=1}^{K(i)} C_{i,k}$. Let us set

$$\Lambda_{i,k} = (C_{i,k} \times C) \cap \Lambda_i, \quad \mu_{i,k}^+ = X_\sharp^+(\pi_0 \mathbb{1}_{\Lambda_{i,k}}) \text{ and } \mu_{i,k}^- = X_\sharp^-(\pi_0 \mathbb{1}_{\Lambda_{i,k}}).$$

We have just cut μ^+ and μ^- into pieces. Let us call informally μ_i^+ the pieces of μ^+ for which the Wasserstein distance to the corresponding part μ_i^- of μ^- is of order $2^i \frac{w}{2}$. Then $\mu_{i,k}^+$ is the part of μ_i^+ whose support is in the cube $C_{i,k}$. What we have now gained is that each $\mu_{i,k}^+$ has a specified diameter of order $2^i w$ and is at a distance to its corresponding $\mu_{i,k}^-$ which is of the same order $2^i w$ (see picture 6.2). Let us be a bit more precise. The support of $\mu_{i,k}^+$ is a cube with edge $(2^i - 1)w$. By definition of Λ_i, the maximum distance of a point of $\mu_{i,k}^-$ to a point of $\mu_{i,k}^+$ is less than $(2^{i+1} - 1)\frac{w}{2}$. Thus the supports of $\mu_{i,k}^-$ and $\mu_{i,k}^+$ are both contained in a same cube with edge $6 \cdot 2^i w$.

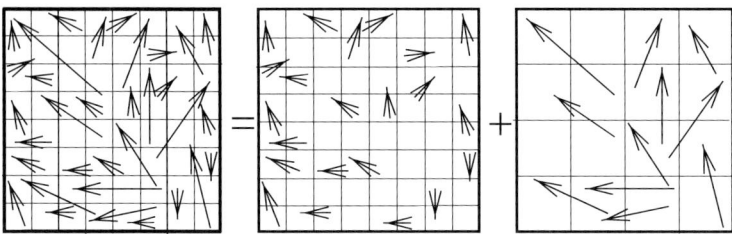

Fig. 6.2. Decomposition of Monge's transportation into the sets $\Lambda_{i,k}$.

By the scaling property of the E^α distance (6.3), we deduce that for some constant c, depending only on α and N, holds:

$$E^\alpha(\mu_{i,k}^+, \mu_{i,k}^-) \leq c 2^i w \pi_0 (\Lambda_{i,k})^\alpha.$$

From this last relation, the sub-additivity of E^α, Hölder inequality, (6.4) and the bound on $K(i)$ given in (6.5), one obtains in turn

$$E^\alpha(\mu^+, \mu^-) \leq \sum_{i,k} E^\alpha(\mu_{i,k}^+, \mu_{i,k}^-)$$

$$\leq \sum_{i,k} c 2^i w \pi_0(\Lambda_{i,k})^\alpha = c \sum_{i,k} (2^i w \pi_0(\Lambda_{i,k}))^\alpha (2^i w)^{1-\alpha}$$

$$\leq c \left(\sum_{i,k} (2^i w \pi_0(\Lambda_{i,k})) \right)^\alpha \left(\sum_{i,k} 2^i w \right)^{1-\alpha}$$

$$\leq c \left(\sum_i (2^i w \pi_0(\Lambda_i)) \right)^\alpha \left(\sum_{i=0}^{I} K(i) 2^i w \right)^{1-\alpha}$$

$$\leq c w^\alpha \left(\sum_{i=0}^{I} \left(\frac{c}{(2^{i+1}-1)w} \right)^N 2^i w \right)^{1-\alpha}$$

$$\leq c w^{\alpha + (1-N)(1-\alpha)} \left(\sum_{i=0}^{I} 2^{i(1-N)} \right)^{1-\alpha}$$

$$\leq c w^{\alpha N - (N-1)} = c W_1(\mu^+, \mu^-)^{\alpha N - (N-1)},$$

where c denotes various constants depending only on N and α and where the last two inequalities are valid if $N \geq 2$ so that the series $\sum_{i=0}^{\infty} 2^{i(1-N)}$ is convergent.

In the case $N = 1$ a different proof is needed. In this case we know how does an optimal transportation for $E^\alpha(\mu^+, \mu^-)$ look like. In the one-dimensional setting, we have

$$E^\alpha(\mu^+, \mu^-) = \int_0^1 |\theta(x)|^\alpha dx.$$

The function θ plays the role of the multiplicity and it is given by

$$\theta(x) = \mu([0, x]), \quad \mu := \mu^+ - \mu^-,$$

as a consequence of its constraint on the derivative. Hence we have

$$E^\alpha(\mu^+, \mu^-) = \int_0^1 |\mu([0, x])|^\alpha dx \leq \left[\int_0^1 |\mu([0, x])| dx \right]^\alpha,$$

where the inequality comes from Jensen's inequality. Then we set

$$A = \{x \in [0, 1] : \mu([0, x]) > 0\}$$

and $h(x) = \mathbb{1}_A(x) - \mathbb{1}_{[0,1] \setminus A}(x)$ and we have

$$\int_0^1 |\mu([0, x])| dx = \int_0^1 \mu([0, x]) h(x) dx = \int_0^1 h(x) dx \int_0^1 \mathbb{1}\{t \leq x\} \mu(dt)$$

$$= \int_0^1 \mu(dt) \int_t^1 h(x) dx = \int_0^1 u(t) \mu(dt) \leq W_1(\mu^+, \mu^-),$$

where $u(t) = \int_t^1 h(x) dx$ is a Lipschitz continuous function whose Lipschitz constant does not exceed 1 as a consequence of $|h(x)| \leq 1$. Thus the last inequality is justified by the duality formula (see [86], Theorem 1.14, page 34):

$$W_1(\mu^+, \mu^-) = \sup_{v \in Lip_1} \int_0^1 v \, d(\mu^+ - \mu^-).$$

Hence it follows easily $E^\alpha(\mu^+, \mu^-) \leq W_1(\mu^+, \mu^-)^\alpha$, which is the thesis for the one dimensional case.

Remark 6.17. The assumption $\alpha > 1 - 1/N$ cannot be removed since, for $N \geq 2$, if we remove this assumption, the quantity E^α could be infinite while W_1 is always finite. In dimension 1 the only uncovered case is $\alpha = 0$. In this case E^α is in fact always finite but, for instance if $\mu^+ = w_0$ and $\mu^- = (1 - \varepsilon) w_0 + \varepsilon w_1$ one has $E^\alpha(\mu^+, \mu^-) = 1$ while $W_1(\mu^+, \mu^-) = \varepsilon$. As ε is as small as we want, this excludes any desired inequality.

Remark 6.18. The exponent $N(\alpha - (1 - 1/N))$ cannot be improved as can be seen from the following example.

Example 6.19. There exists a sequence (μ_n^+, μ_n^-) of pairs of probability measures on the cube C such that

$$E^\alpha(\mu_n^+, \mu_n^-) = cn^{-N(\alpha - (1 - 1/N))} \text{ and } W_1(\mu_n^+, \mu_n^-) = c/n.$$

Proof. It is sufficient to divide the cube C into n^N small cubes of edge $1/n$ and to set $\mu_n^+ = \sum_{i=1}^{n^N} \frac{1}{n^N} \delta_{x_i}$ and $\mu_n^- = \sum_{i=1}^{n^N} \frac{1}{n^N} \delta_{y_i}$, where each x_i is a vertex of one of the n^N cubes (let us say the one with minimal sum of the N coordinates) and the corresponding y_i is the center of the same cube. In this way y_i realizes the minimal distance to x_i among the y_j's. Thus the optimal configuration both for E^α and W_1 is given by linking any x_i directly to the corresponding y_i. In this way we have

$$E^\alpha(\mu_n^+, \mu_n^-) = n^N \left(\frac{1}{n^N} \right)^\alpha \frac{c}{n} = cn^{-N(\alpha - (1 - 1/N))}$$

$$W_1(\mu_n^+, \mu_n^-) = n^N \frac{1}{n^N} \frac{c}{n} = \frac{c}{n},$$

where $c = \frac{\sqrt{N}}{2}$.

Remark 6.20. One can deduce easily inequalities between E^α and W_p by using standard inequalities between W_1 and W_p, namely $cW_p^p \leq E^\alpha \leq cW_p^{N(\alpha - (1 - 1/N))}$. The right hand inequality is sharp by using again example 6.19. It is not clear instead whether the left-hand inequality is optimal.

Remark 6.21. Since the W_1 distance between two probability measures is always finite, Proposition 6.16 gives another proof of the fact that there is a traffic plan at finite cost for $\alpha \in (1 - \frac{1}{N}, 1]$.

7 The Tree Structure of Optimal Traffic Plans and their Approximation

We shall get in this chapter a deeper insight of the structure of optimal traffic plans. Section 7.1 proves a *single path property* for optimal traffic plans: almost all fibers passing from x to y follow the same path between x and y. By a slight modification of the optimal traffic plan, this statement can be made strict: All fibers (not just almost all) passing by x and y follow the same path. Section 7.2 extends the single path property in the case of optimal traffic plans for the irrigation problem. In that case optimal traffic plans have no circuits, in other terms are *trees*. Section 7.3 uses the single path property to show that optimal traffic plans can be monotonically approximated by finite irrigation graphs. The steps leading to this conclusion have their own interest. In particular it is shown that any optimal traffic plan is a finite or countable union of *trunk trees*, namely trees whose fibers pass all by some point. The presentation here is inspired from [13]. Several techniques come from papers by Xia [94] and Maddalena-Solimini [58]. The proof of the bi-Lipschitz regularity of fibers with positive flow follows [95] and the pruning and theorem is borrowed from Devillanova and Solimini [79]. The monotone approximation theorem has also been proved in [58].

7.1 The Single Path Property

In all this chapter as well as the rest of the book, we will use the term fiber in place of essential fiber, without loss of generality. In other words, we will always suppose that fibers are considered to have a positive multiplicity $\mathcal{H}^1-$almost everywhere on the path they define (see Definition 3.38).

Definition 7.1. *Let χ be a loop-free traffic plan, so that $t_x(\omega) := \chi^{-1}(\omega, \cdot)(x)$ is well defined. Let x, y in S_χ. Define*

$$\Omega_{\overrightarrow{xy}} := \{\omega \in \Omega_x^\chi \cap \Omega_y^\chi : t_x(\omega) < t_y(\omega)\},$$

the set of fibers passing through x and then through y. The restriction of χ to $\cup_{\omega \in \Omega_{\overrightarrow{xy}}} \{\omega\} \times [t_x(\omega), t_y(\omega)]$ is denoted by χ_{xy}. It is the traffic plan made of all pieces of fibers of χ joining x to y. Denote its support by $\Gamma^{xy} = S_{\chi_{xy}}$.

Throughout this Chapter we assume that χ be a loop-free traffic plan. In particular, χ has simple paths and by Proposition 4.8 we have $E^\alpha(\chi) = \mathcal{E}^\alpha(\chi)$.

M. Bernot et al., *Optimal Transportation Networks*. Lecture Notes in Mathematics 1955.
© Springer-Verlag Berlin Heidelberg 2009

Lemma 7.2. *Let $\alpha \in [0,1)$ and χ be an optimal traffic plan (either for the irrigation or the who goes where problem) such that $E^\alpha(\chi) < \infty$. Let x, y be such that $\Omega_{\overrightarrow{xy}}$ has positive measure. Set for every disjoint pair Ω_1, $\Omega_2 \subset \Omega_{\overrightarrow{xy}}$ with $|\Omega_1|$, $|\Omega_2| > 0$, $\theta_i(z) := |\Omega_i \cap \Omega_z|$ for $i = 1, 2$. Then $(\theta_1(z) - \theta_2(z)\frac{|\Omega_1|}{|\Omega_2|}) = 0$ $\mathcal{H}^1$-almost everywhere on Γ^{xy}.*

Proof. For z in Γ^{xy} write $\bar{\theta}(z) = |\Omega_z \setminus (\Omega_1 \cup \Omega_2))|$. Notice that the multiplicity of χ at z is $|z|_\chi = \theta_1(z) + \theta_2(z) + \bar{\theta}(z)$ for all $z \in \mathbb{R}^N$. Assume by contradiction that $(\theta_1(z) - \theta_2(z)\frac{|\Omega_1|}{|\Omega_2|}) \neq 0$ on a subset of Γ^{xy} with positive $\mathcal{H}^1$ measure. To obtain a contradiction with the optimality of χ, we are going to prove that either transferring mass conveyed by $\chi_{|\Omega_1}$ between x and y to mass conveyed by $\chi_{|\Omega_2}$ or conversely decreases the energy of the traffic plan (see Figure 7.1). Let ρ be the proportion of fibers of Ω_2 to be transferred to Ω_1. Set

$$\theta^\rho(z) := (1 + \rho)\theta_1(z) + (1 - \rho\frac{|\Omega_1|}{|\Omega_2|})\theta_2(z) + \bar{\theta}(z).$$

Take $\rho\frac{|\Omega_1|}{|\Omega_2|} \leq \frac{1}{2}$ so that

$$\frac{1}{2}|z|_\chi \leq \theta^\rho(z). \tag{7.1}$$

Let us prove that there exists a traffic plan χ_ρ with the same transference plan as χ and such that $|z|_{\chi_\rho} = \theta^\rho(z)$. Up to a measure preserving bijection, we can suppose for the sake of convenience that $\Omega_1 = [0, |\Omega_1|]$ and $\Omega_2 =]|\Omega_1|, |\Omega_2| + |\Omega_1|]$. Let us denote $\tilde{\Omega}_1 = [0, |\Omega_1|(1 + \rho)]$ and $\tilde{\Omega}_2 =]|\Omega_1|(1 + \rho), |\Omega_2| + |\Omega_1|]$. The application

$$\psi(\omega) = \begin{cases} \frac{|\Omega_1|}{|\tilde{\Omega}_1|}\omega & \text{if } \omega \in \tilde{\Omega}_1 \\ \frac{|\Omega_2|}{|\tilde{\Omega}_2|}(\omega - |\tilde{\Omega}_1|) + |\Omega_1| & \text{if } \omega \in \tilde{\Omega}_2 \\ \omega & \text{if } \omega \in \Omega \setminus (\tilde{\Omega}_1 \cup \tilde{\Omega}_2) \end{cases}$$

contracts $\tilde{\Omega}_1$ onto Ω_1 and dilates $\tilde{\Omega}_2$ onto Ω_2 (see Figure 7.2). Set $|I_{\psi(\omega)}| = t_y(\psi(\omega)) - t_x(\psi(\omega))$ and define

$$\chi_\rho(\omega, t) = \begin{cases} \chi(\omega, t) & \text{if } t \leq t_x(\omega) \\ \chi(\psi(\omega), t - t_x(\omega) + t_x(\psi(\omega))) & \text{if } t \in [t_x(\omega), t_x(\omega) + |I_{\psi(\omega)}|] \\ \chi(\omega, t - t_x(\omega) - |I_{\psi(\omega)}| + t_y(\omega)) & \text{if } t > t_x(\omega) + |I_{\psi(\omega)}| \end{cases}$$

which is obtained by transferring uniformly mass of Ω_2 onto Ω_1 for the fibers joining x to y. The traffic plan χ_ρ is by definition such that $|z|_{\chi_\rho} = \theta^\rho(z)$. Further, the transference plan of χ_ρ is the same as the one of χ since $\chi_\rho(\omega, 0) = \chi(\omega, 0)$ and $\chi_\rho(\omega, \infty) = \chi(\omega, \infty)$ for all $\omega \in [0, 1]$. Let us compare the costs of χ and χ_ρ. Define the balance of the energy as

$$f(\rho) = E^\alpha(\chi_\rho) - E^\alpha(\chi).$$

By definition of E^α,

$$f(\rho) = \int_{\Gamma^{xy}} (\theta^\rho(z)^\alpha - |z|_\chi^\alpha) d\mathcal{H}^1.$$

Thus

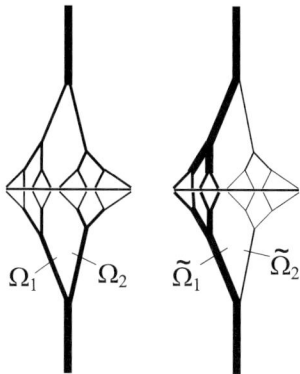

Fig. 7.1. Illustrates the proof of Lemma 7.2. Suppose that in an optimal traffic plan χ two sets of fibers Ω_1 and Ω_2 go from x to y. Some part of the mass of Ω_2 can be conveyed through the fibers of Ω_1, or conversely, without changing the irrigated measures or the transference plan of χ . Thus the modified traffic plan displayed on the right has an energy larger or equal to the energy of χ.

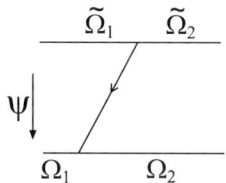

Fig. 7.2. Illustrates the proof of Lemma 7.2. The application ψ is contracting $\tilde{\Omega}_1$ on Ω_1 and dilating $\tilde{\Omega}_2$ on Ω_2.

$$f'(\rho) = \alpha \int_{\Gamma^{xy}} \theta^\rho(z)^{\alpha-1}(\theta_1(z) - \theta_2(z)\frac{|\Omega_1|}{|\Omega_2|}) d\mathcal{H}^1,$$

and

$$f''(\rho) = \alpha(\alpha-1) \int_{\Gamma^{xy}} \theta^\rho(z)^{\alpha-2}(\theta_1(z) - \theta_2(z)\frac{|\Omega_1|}{|\Omega_2|})^2 d\mathcal{H}^1.$$

Notice that the three above integrals and both derivations are licit because of (7.1) and $\theta_1(z), \theta_2(z) \leq |z|_\chi$.

By assumption, $(\theta_1(z) - \theta_2(z)\frac{|\Omega_2|}{|\Omega_1|}) \neq 0$ on a subset of Γ^{xy} with positive $\mathcal{H}^1$ measure. Since $\alpha < 1$, we obtain that $f''(\rho) < 0$. Thus $f'(\rho) < f'(0) = \alpha \int_{\Gamma^{xy}} (|z|_\chi^{\alpha-1}(\theta_1(z) - \theta_2(z)\frac{|\Omega_2|}{|\Omega_1|}) d\mathcal{H}^1$. Without loss of generality assume that $f'(0) \leq 0$, otherwise one can exchange Ω_1 and Ω_2. Thus $f'(\rho) < 0$ and $f(\rho) < f(0) = 0$ for a sufficiently small ρ. This inequality contradicts the optimality of χ.

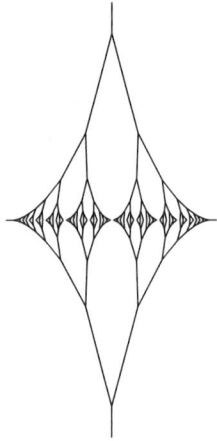

Fig. 7.3. This traffic plan is obtained through the concatenation of a traffic plan transporting a Dirac mass to the Lebesgue measure on a segment and a traffic plan transporting the Lebesgue measure on a segment to a Dirac mass. Proposition 7.4 proves that such a structure is not optimal.

Definition 7.3. *A traffic plan χ has the* single path property *if for every pair x, y such that $|\Omega_{\overrightarrow{xy}}| > 0$, almost all fibers in $\Omega_{\overrightarrow{xy}}$ coincide between x and y with an arc Γ^{xy} joining x to y. We say that the traffic plan has the* strict single path property *if for every x, y, either there is no fiber joining x to y or $|\Omega_{\overrightarrow{xy}}| > 0$ and all fibers in Γ^{xy} coincide between x and y.*

Proposition 7.4. (Single path property) *Let $\alpha \in [0, 1)$ and χ be an optimal traffic plan (either for the irrigation or the who goes where problem) such that $E^\alpha(\chi) < \infty$. Then χ is single path.*

Proof. We refer to Figure 7.3 for an example of configuration excluded by Proposition 7.4. In case that

$$|z|_{\chi_{xy}} = |\Omega_{\overrightarrow{xy}}| \qquad \mathcal{H}^1\text{-a.e. } z \in \Gamma^{xy}, \tag{7.2}$$

we have by Fubini's theorem

$$\mathcal{H}^1(\Gamma^{xy})|\Omega_{\overrightarrow{xy}}| = \int_{\Gamma^{xy}} |z|_{\chi_{xy}}\, d\mathcal{H}^1(z) = \int_{\Gamma^{xy}} \int_{\Omega_{\overrightarrow{xy}}} \mathbb{1}_{z \in \mathrm{Im}(\chi_{xy}(\omega))}\, d\omega\, d\mathcal{H}^1(z)$$

$$= \int_{\Omega_{\overrightarrow{xy}}} \mathcal{H}^1(\Gamma^{xy} \cap \mathrm{Im}\chi(\omega))\, d\omega.$$

Hence

$$\mathcal{H}^1(\Gamma^{xy} \cap \mathrm{Im}\chi(\omega)) = \mathcal{H}^1(\Gamma^{xy}) \qquad \text{a.e. } \omega \in \Omega_{\overrightarrow{xy}}.$$

This implies that $\Gamma^{xy} = \mathrm{Im}\,\chi(\omega)$ (mod $\mathcal{H}^1$) for almost all $\omega \in \Omega_{\overrightarrow{xy}}$ and the statement is proved. If instead $|z|_{\chi_{xy}} < |\Omega_{\overrightarrow{xy}}|$ on a subset of Γ^{xy} with positive $\mathcal{H}^1$ measure, the conclusion follows immediately from Lemma 7.2 above and Lemma 7.5 below.

Lemma 7.5. *Let* χ *a traffic plan,* $x \neq y$ *in the support of* χ, $\Omega_{\overrightarrow{xy}}$ *and* Γ^{xy} *as defined above. Assume that*

$$\mathcal{H}^1(\{z \in \Gamma^{xy}, |z|_{\chi_{xy}} < |\Omega_{\overrightarrow{xy}}|\}) > 0.$$

Then there are two disjoint sets Ω_1, $\Omega_2 \subset \Omega_{\overrightarrow{xy}}$ *such that* $\theta_1(z) - \theta_2(z)\frac{|\Omega_1|}{|\Omega_2|} > 0$ *on a subset of* Γ^{xy} *with positive* $\mathcal{H}^1$ *measure.*

Proof. Given two sets E and F denote $E \Delta F = (E \setminus F) \cup (F \setminus E)$. Fix $a > 0$ small enough so that $\mathcal{H}^1(L) > 0$ where $L = \{z \in X : 2a < |z|_{\chi_{xy}} < |\Omega_{\overrightarrow{xy}}| - a\}$. Without loss of generality we can assume that $\Omega_{\overrightarrow{xy}}$ is a measurable subset of $I = [0, |\Omega|]$. Consider an enumeration $(O_n)_{n \in \mathbb{N}}$ of all finite unions of intervals of I with rational endpoints. Then for every measurable subset E of I and every $\varepsilon > 0$ there is n such that $|E \Delta O_n| < \varepsilon$. Let

$$L_n := \{z \in L, |O_n \Delta(\Omega_z \cap \Omega_{\overrightarrow{xy}})| < \varepsilon\}.$$

By Lemma 7.6 below each L_n is a borelian. Since $\cup_n L_n = L$, $\mathcal{H}^1(L_k) > 0$ for some k. By the triangular inequality, for every $z, z' \in L_k$,

$$|(\Omega_z \Delta \Omega_{z'}) \cap \Omega_{\overrightarrow{xy}}| < 2\varepsilon. \qquad (7.3)$$

Fix $z_1 \in L_k$ and take $\Omega_1 = \Omega_{z_1} \cap \Omega_{\overrightarrow{xy}}$ and $\Omega_2 = \Omega_{\overrightarrow{xy}} \setminus \Omega_1$ so that $|\Omega_2| > a$. Thus $\theta_1(z_1) = |\Omega_1|$ and $\theta_2(z_1) = 0$. For every $z \in L_k$ the inequality (7.3) yields

$$|\theta_1(z) - \theta_1(z_1)| = |\theta_1(z) - |\Omega_1|| < 2\varepsilon,$$
$$|\theta_2(z) - \theta_2(z_1)| = |\theta_2(z)| < 2\varepsilon.$$

These last relations imply

$$\theta_1(z) - \theta_2(z)\frac{|\Omega_1|}{|\Omega_2|} \geq |\Omega_1| - 2\varepsilon - 2\varepsilon\frac{|\Omega_1|}{|\Omega_2|} \geq 2a - 2\varepsilon - \frac{2\varepsilon|\Omega_1|}{a} > 0$$

provided ε is taken small enough.

Lemma 7.6. *Let χ be a traffic plan and $O, U \subset \Omega$. Then the function $z \in \mathbb{R}^N \mapsto |(\Omega_z \cap U) \Delta O|$ is borelian.*

Proof. Observe that $|(\Omega_z \cap U) \Delta O| = |\Omega_z \cap U| + |O| - 2|\Omega_z \cap U \cap O|$ and that the functions $z \mapsto |\Omega_z \cap U|$ and $z \mapsto |\Omega_z \cap U \cap O|$ are upper semicontinuous. Let us prove this property, actually shown in [59]. If $z_n \to z$ and if $\omega \in \Omega_{z_n}$ infinitely often, then $\omega \in \Omega_z$ by continuity of each fiber. Thus $\limsup_n \Omega_{z_n} \cap U \subset \Omega_z \cap U$ and therefore $|\Omega_z \cap U| \geq \limsup_n |\Omega_{z_n} \cap U|$. Thus $z \to |\Omega_z \cap U|$ is upper semicontinuous and therefore borelian.

7.2 The Tree Property

We now aim at a more powerful property for optimal traffic plans for the irrigation problem. They have no circuit and in consequence have tree structure. This property won't be valid for optimal traffic plans for the who goes where problem.

Lemma 7.7. *Let χ be an optimal traffic plan such that $E^\alpha(\chi) < \infty$. Let x, y be in the support of χ such that $|\{\omega : x, y \in \chi(\omega, \mathbb{R}^+)\}| > 0$. Then, either $|\Omega_{\overrightarrow{xy}}| = 0$ or $|\Omega_{\overrightarrow{yx}}| = 0$. We denote Ω_{xy}, the one that has positive measure and Γ_{xy} the only arc between x and y, i.e. Γ^{xy} or Γ^{yx} whether $|\Omega_{\overrightarrow{xy}}| \neq 0$ or $|\Omega_{\overrightarrow{yx}}| \neq 0$.*

Proof. If $\Omega_{\overrightarrow{xy}}$ and $\Omega_{\overrightarrow{yx}}$ are both non negligible, consider two subsets of same positive measure $\Omega_1 \subset \Omega_{\overrightarrow{xy}}$ and $\Omega_2 \subset \Omega_{\overrightarrow{yx}}$ and $\phi : \Omega_1 \to \Omega_2$ bijective and measure preserving. Let us define $\tilde{\chi}$ as χ for all $\omega \notin \Omega_1 \cup \Omega_2$. For all $\omega \in \Omega_1$, we define

$$\tilde{\chi}(\omega, t) = \begin{cases} \chi(\omega, t) & \text{if } t \leq t_x(\omega) \\ \chi(\phi(\omega), t - t_x(\omega) + t_x(\phi(\omega))) & \text{if } t \geq t_x(\omega) \end{cases}$$

We define $\tilde{\chi}$ in a similar way on Ω_2 (replacing t_x by t_y and ϕ by ϕ^{-1}). The resulting traffic plan $\tilde{\chi}$ is in $\text{TP}(\mu^+(\chi), \mu^-(\chi))$ and has a strictly lower energy than χ. This contradiction proves the lemma.

Proposition 7.8. (No circuits) *[94, Proposition 2.1] Let $\alpha \in [0, 1)$ and let χ be a traffic plan such that $E^\alpha(\chi) < \infty$. If there is a circuit in χ, namely points $(x_i)_{i=1}^n$ in the support of χ such that $|\Omega_{x_i x_{i+1}}| \neq 0$ for all $i \in \{1, \ldots, n-1\}$ and $|\Omega_{x_1 x_n}| \neq 0$, then χ is not optimal for the irrigation problem.*

Proof. By Lemma 7.7, it is consistent to define respectively L^+ and L^- as the set of indices such that respectively $|\Omega_{\overrightarrow{x_i x_{i+1}}}| \neq 0$ and $|\Omega_{\overrightarrow{x_{i+1} x_i}}| \neq 0$. Set $m > 0$ small enough so that we may consider sets $\Omega_i \subset \Omega_{\overrightarrow{x_i x_{i+1}}}$ with $|\Omega_i| = m$ for all i. It is indifferent to prove either that χ or the reversed time traffic plan obtained from χ is not optimal, thus we can assume without loss of generality that

$$\sum_{i \in L^+} \int_0^{l_i} \theta_i(s)^{\alpha-1} ds \leq \sum_{i \in L^-} \int_0^{l_i} \theta_i(s)^{\alpha-1} ds,$$

where $\theta_i(s) = |\gamma_i(s)|_\chi$ and γ_i is an arc parameterization of $\Gamma_{x_i x_{i+1}}$ with length l_i. We now define χ_ε such that each flow along an L^+ arc is increased by $\varepsilon < m$ and each flow along an L^- arc is decreased by ε (see Figure 7.4). This traffic plan can be obtained through a convenient cut and paste of fibers by the same concatenation technique as in Lemma 7.7. The new irrigating and irrigated measures are the same as those of χ. Let us set $f(\varepsilon) = E^\alpha(\chi_\varepsilon) - E^\alpha(\chi)$. We have

$$f(\varepsilon) = \sum_{i \in L^+} \int_0^{l_i} (\theta_i(s) + \varepsilon)^\alpha ds + \sum_{i \in L^-} \int_0^{l_i} (\theta_i(s) - \varepsilon)^\alpha ds - E^\alpha(\chi).$$

Since $\alpha \in [0, 1)$, the function f is strictly concave and

$$f'(\varepsilon) < f'(0) = \sum_{i \in L^+} \int_0^{l_i} (\theta_i(s))^{\alpha-1} ds - \sum_{i \in L^-} \int_0^{l_i} (\theta_i(s))^{\alpha-1} ds \leq 0.$$

Thus $f(\varepsilon) < f(0) = 0$, the cost of χ_ε is lower than the cost of χ and χ is not optimal.

The non existence of circuits in optima of irrigation problems has been stated in many papers addressing the problem from different points of view [58, 70, 94] as reviewed in Chapter 9. Similar properties for optimal transportation networks for urban planning have been proved in [22, 24].

Corollary 7.9. *Suppose that P is an optimal traffic plan, that two curves $\gamma_0, \gamma_1 \in K_{arc} \cap K_{inj}$ meet twice (i.e. $\gamma_0(s_0) = \gamma_1(s_1)$, $\gamma_0(t_0) = \gamma_1(t_1)$ and $s_i \neq t_i$) and that $|\gamma_0(t)|_P \geq c > 0$ for any $t \in [s_0, t_0]$. Then either both curves coincide in the trajectory between the two common points or we have $\int_{s_0}^{t_0} |\gamma_0(t)|_P^{\alpha-1} dt < \int_{s_1}^{t_1} |\gamma_1(t)|_P^{\alpha-1} dt$.*

Proof. This is a straightforward adaptation of the preceding proof to the circuit made of the arcs $\gamma_0([s_0, t_0])$ and $\gamma_1([s_1, t_1])$: if both curves do not coincide and the integrals are not ordered as indicated one can transfer mass from γ_0 to γ_1 and save energy by strict concavity, getting a contradiction.

7.3 Decomposition into Trees and Finite Graphs Approximation

Definition 7.10. *A traffic plan satisfying the single path property and such that all fibers pass by some $x \in X$ will be called a trunk tree. If $\mu^+(\chi)$ is a*

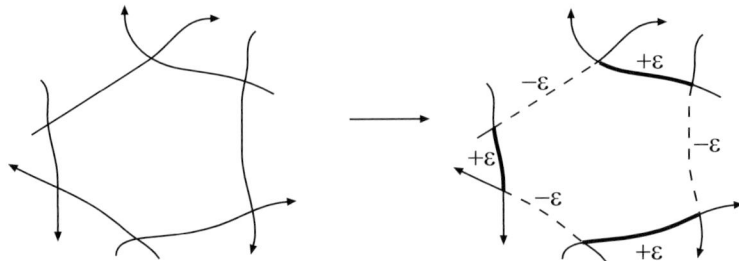

Fig. 7.4. Illustrates the proof of Proposition 7.8. The modification of the traffic plan χ consists in transfering a multiplicity ε from all arcs Γ_j where $j \in L^-$ to arcs Γ_i where $i \in L^+$. This perturbation gives a new traffic plan χ_ε which has a lower cost than χ and the same irrigating and irrigated measures (but not the same transference plan).

Dirac mass the trunk tree is called a pattern. *A finite irrigation graph* denotes *any traffic plan having the strict single path property which is the union of a finite set of single paths with constant multiplicities. A* finite irrigation tree *is any finite irrigation graph which is a trunk tree. In both cases the irrigation measure is finite atomic.*

Proposition 7.11. *Let χ be a trunk tree with finite energy. Assume that χ is parameterized by arc-length. Then there is a sequence ξ_n of restrictions of χ such that:*

- *each ξ_n is a finite irrigation tree;*
- $\mathcal{D}(\xi_n) \subset \mathcal{D}(\xi_{n+1})$, $T_{\xi_n}(\omega) \to T_\chi(\omega)$ *a.e.;*
- $E^\alpha(\xi_n) \to E^\alpha(\chi)$;
- $\mu^+(\xi_n) \to \mu^+(\chi)$, $\mu^-(\xi_n) \to \mu^-(\chi)$;

Assume that all fibers of χ pass by the point x. Set $t_x(\omega) = \chi_\omega^{-1}(x)$. For almost every ω the function $t \mapsto |\chi(\omega, t)|_\chi$ is nondecreasing on $[0, t_x(\omega)[$ and nonincreasing on $]t_x(\omega), +\infty]$. In consequence, for all $0 < a < b < T_\chi(\omega)$ one has $\inf_{t\in[a,b]} |\chi(\omega, t)|_\chi > 0$. Finally, by changing Ω in a zero measure set, the restriction ξ of χ to the domain $\{(\omega, t) : 0 < t < T_\chi(\omega)\}$ is strictly single path.

Proof. We can consider χ as the concatenation of a single trunk tree from μ^+ to δ_x and another one from δ_x to μ^-. Thus it is enough to prove the above properties for (e.g.) a tree starting at x and irrigating μ^-. Concatenating the approximating trees after and before x will give the sought for sequence ξ_n.

All considered paths and fibers are parameterized by length. In the following we shall consider a countable set of properties satisfied for almost every $\omega \in \Omega$. Every time, we shall remove from Ω the fibers not satisfying the property. In that way all considered properties will become true for every fiber of some $\tilde{\Omega}$ with $|\tilde{\Omega}| = |\Omega|$. For simplicity the resulting set of fibers $\tilde{\Omega}$ will still be called Ω.

For simplicity, let us write $t(y)$ instead of $t_x(y)$, $y \in S_\chi$. By the single path property, for any $y \in S_\chi$, there is a path $\gamma_y : [0, t(y)] \to X$ joining x to y such that for almost every $\omega \in \Omega_{xy}$ and $0 \le s \le t(y)$, $\chi(\omega, s) = \gamma_y(s)$. Let y_1 satisfying

$$|\Omega_{xy_1}| t(y_1) \ge \frac{1}{2} \max_{y \in S_\chi} |\Omega_{xy}| t(y) > 0$$

and set $\xi_1 = \chi_{|\Omega_{xy_1} \times [0, t(y_1)[}$, where Ω_{xy_1} contains only the fibers coinciding with γ_{y_1} on $[0, t(y_1)]$. Let now y be such that $|y|_{\xi_1} < |y|_\chi$. By the single path property, we are in one of the following situations:

1. Almost all fibers of Ω_{xy} pass by y_1 and then $\Omega_{xy} \subset \Omega_{xy_1}$ almost everywhere. Then set $s_2(y) = \gamma_y^{-1}(y_1)$ and $\Omega_2 = \Omega_{xy}$
2. Almost all fibers of Ω_{xy_1} pass by y and then $\Omega_{xy_1} \subset \Omega_{xy}$ almost everywhere. In that case set $s_2(y) = 0$ and $\Omega_2 = \Omega_{xy} \setminus \Omega_{xy_1}$.
3. The sets Ω_{xy} and Ω_{xy_1} are disjoint almost everywhere. Set $s_2(y) = 0$ and $\Omega_2 = \Omega_{xy}$.

Let y_2 satisfying

$$(|y_2|_\chi - |y_2|_{\xi_1})(t(y_2) - s_2(y_2)) \ge \frac{1}{2} \max_{y, |y|_{\xi_1} < |y|_\chi} (|y|_\chi - |y|_{\xi_1})(t(y) - s_2(y)) > 0.$$

Define ξ_2 by the concatenation of ξ_1 and $\chi_{|\Omega_2 \times [s_2(y_2), t(y_2)[}$ if $s_2(y) \ne 0$ or by the union of ξ_1 and $\chi_{|\Omega_2 \times [0, t(y_2)[}$ otherwise. (Again here the exceptional fibers of Ω_2 have been removed from Ω.)

One can iterate the construction. At each step ξ_n is the union or the concatenation of ξ_{n-1} with the constant multiplicity path $\chi_{|\Omega_n \times [s_n(y_n), t(y_n)[}$. One has $s_n(y_n) = \gamma_{y_n}^{-1}(y_k)$ if γ_{y_n} passes by some y_k, $k \le n-1$, and $s_n(y_n) = 0$ otherwise. Thus ξ_n is a finite subtree of χ and

$$0 < \frac{1}{2} \max_{y : |y|_{\xi_{n-1}} < |y|_\chi} (|y|_\chi - |y|_{\xi_{n-1}})(t(y) - s_n(y)) \tag{7.4}$$

$$\le (|y_n|_\chi - |y_n|_{\xi_{n-1}})(t(y_n) - s_n(y_n)). \tag{7.5}$$

Notice that the sets $(\Omega_{y_n}^\chi \setminus \Omega_{y_n}^{\xi_{n-1}}) \times [s_n(y_n), t(y_n)[$ are a.e. disjoint. Hence,

$$\sum_n (|y_n|_\chi - |y_n|_{\xi_{n-1}})(t(y_n) - s_n(y_n)) \le \int_\Omega T_\chi(\omega) < \infty$$

and by (7.4) as $n \to \infty$,

$$\max_{y : |y|_{\xi_{n-1}} < |y|_\chi} (|y|_\chi - |y|_{\xi_{n-1}})(t(y) - s_n(y)) \to 0. \tag{7.6}$$

Define a limit traffic plan ξ by $\xi(\omega, t) = \xi_n(\omega, t)$ if $(\omega, t) \in \mathcal{D}(\xi_n)$. By construction ξ is the restriction of χ to the set $\cup_{\omega \in \Omega} \{\omega\} \times [0, T_\xi(\omega)[$ with $T_\xi(\omega) = \sup_{n, \omega \in \Omega_{xy_n}} t(y_n)$. Let us show that $T_\xi(\omega) = T_\chi(\omega)$ almost everywhere.

If for some y in the support of χ one has $|y|_\xi < |y|_\chi$, the relation (7.6) yields $t(y) - s_n(y) \to 0$. If $s_n(y) = 0$ for every n, then $t(y) = 0$ and therefore $y = x$ which is impossible since χ contains no zero length path (see Section 5.3.1). Hence, there is an increasing subsequence n_k such that $s_{n_k}(y) \to t(y)$. Consequently the path $\gamma_y(s), 0 \le s < t(y)$ with multiplicity $|y|_\chi$ is a restriction of ξ. Thus ξ irrigates the atomic mass $|y|_\chi \delta_y$. This implies that the set of points y such that $|y|_\xi < |y|_\chi$ is at most countable. One concludes that $|y|_\xi = |y|_\chi$ $\mathcal{H}^1$-a.e. Using the formula $\int_\Omega T_\chi(\omega)d\omega = \int_X |x|_\chi d\mathcal{H}^1$ and $T_\xi(\omega) \le T_\chi(\omega)$ a.e., one deduces that $T_\chi(\omega) = T_\xi(\omega)$ almost everywhere. By construction the function $t \mapsto |\xi_n(\omega, t)|_{\xi_n}$ is nonincreasing for every ω. This property still holds for ξ. Thus $t \mapsto |\chi(\omega, s)|_\chi$ is nonincreasing on $[0, T_\chi(\omega)[$ as announced. Again by construction all ξ_n have the strict single path property and this property is inherited by ξ.

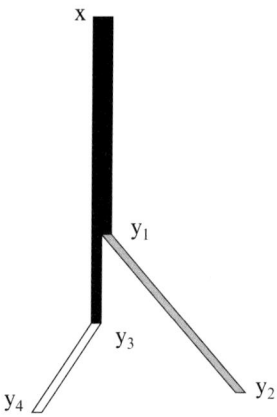

Fig. 7.5. Illustrates the construction of an approximating finite tree in Proposition 7.11. The tree is constructed by adding to an initial path ending at y_1 other paths starting either from the root x or from a former path extremity y_k.

Corollary 7.12 (Pruning Theorem, Solimini-Devillanova [79]). *Let $\varepsilon > 0$, $\alpha \in [0, 1)$ and χ be an optimal trunk tree. Then there exists a finite number n of points $y_i \in S_\chi$ such that, denoting by χ_i the subtree of χ with source point y_i, the χ_i are disjoint and we have*

$$\sum_{i=1}^{n} E^\alpha(\chi_i) < \varepsilon, \tag{7.7}$$

$$(\mu_\chi - \sum_{i=1}^{n} \mu_{\chi_i})(X) < \varepsilon. \tag{7.8}$$

Proof. Without loss of generality, we may assume that χ is parameterized by arc-length. Using the notation and the result of Proposition 7.11, we can take n large enough so that

$$E^\alpha(\chi) - E^\alpha(\xi_n) < \varepsilon \text{ and } |\Omega \setminus \Omega_n| < \varepsilon$$

where Ω_n denotes the set of fibers of ξ_n. The second property can be achieved for n large enough because $T_\chi(\omega) > 0$ a.e. and $T_{\xi_n}(\omega) \to T_\chi(\omega)$ almost everywhere. Let us call $(y_i)_{i=1}^n$ the endpoints of the fibers of ξ_n and let χ_i be the subtree of χ with source point y_i. Then

$$\left(\mu_\chi - \sum_{i=1}^n \mu_{\chi_i}\right)(X) = |\Omega \setminus \Omega_n| < \varepsilon$$

and

$$\sum_{i=1}^n E^\alpha(\chi_i) \leq E^\alpha(\chi) - E^\alpha(\xi_n) < \varepsilon.$$

Lemma 7.13. *Let $\alpha \in [0,1)$ and $\chi : \Omega \times \mathbb{R}^+ \to \mathbb{R}^N$ be a traffic plan with finite energy. Then there is a finite or countable set of points x_j in the support of χ such that $\Omega \subset \cup_j \Omega_{x_j}$. In consequence if χ is optimal it is a finite or countable union of trunk trees.*

Proof. Without loss of generality, assume that χ is parameterized by arc-length. Since $\alpha < 1$ and χ has finite energy, $|\chi(\omega, t)|_\chi > 0$ a.e. in $\Omega \times \mathbb{R}^+$. Choose $x_0 \in \mathrm{Im}\,\chi$ such that $|x_0|_\chi \geq \sup_x |x|_\chi - \varepsilon_0 > 0$ for some $\varepsilon_0 > 0$. Define $\tilde{\Omega}_{x_0} := \Omega_{x_0}$, and $\tilde{\chi}_{x_0} := \chi_{|\tilde{\Omega}_{x_0} \times \mathbb{R}^+}$.

If $|x|_\chi - |x|_{\tilde{\chi}_{x_0}} = 0$ on $\mathrm{Im}\chi$ then χ is equivalent to $\tilde{\chi}_{x_0}$. Indeed, in this case by Lemma 5.2

$$E^\alpha(\chi_{|\Omega \setminus \tilde{\Omega}_{x_0}}) = \int_{\Omega \setminus \tilde{\Omega}_{x_0}} \int_0^\infty |\chi(\omega, t)|_{\chi_{|\Omega \setminus \tilde{\Omega}_{x_0}}}^{\alpha - 1} d\omega dt$$
$$\leq E^\alpha(\chi) < \infty.$$

Since $|\chi(\omega, t)|_{\chi_{|\Omega \setminus \tilde{\Omega}_{x_0}}} = |\chi(\omega, t)|_\chi - |\chi(\omega, t)|_{\tilde{\chi}_{x_0}} = 0$ and since all fibers have positive length this implies $|\Omega \setminus \tilde{\Omega}_{x_0}| = 0$. Thus, we may assume that $|x|_\chi - |x|_{\tilde{\chi}_{x_0}} > 0$ for some point $x \in \mathrm{Im}\,\chi$. Then choose $x_1 \in \mathrm{Im}\,\chi$ such that $|x_1|_\chi - |x_1|_{\tilde{\chi}_{x_0}} \geq \sup_x(|x|_\chi - |x|_{\tilde{\chi}_{x_0}}) - \varepsilon_1 > 0$ for some $0 < \varepsilon_1 < \varepsilon_0/2$. Consider $\tilde{\Omega}_{x_1} := \Omega_{x_1} \setminus \tilde{\Omega}_{x_0}$ and $\tilde{\chi}_{x_1} = \chi_{x_1}|_{\tilde{\Omega}_{x_1} \times \mathbb{R}^+}$.

Assume that proceeding in this way we have constructed $(x_i)_{i=0}^k$. Either $|x|_\chi = \sum_{j=0}^k |x|_{\tilde{\chi}_{x_j}}$ on $\mathrm{Im}\,\chi$, in which case by the same argument as above $\Omega = \cup_{j=0}^k \tilde{\Omega}_{x_j}$ (modulo a null set) and χ is equivalent to $\cup_{j=0}^k \tilde{\chi}_{x_j}$, in the sense of Definition 5.3. Or $|x|_\chi - \sum_{j=0}^k |x|_{\tilde{\chi}_{x_j}} > 0$ for some point of $\mathrm{Im}\,\chi$, in which case we construct x_{k+1} such that $|x_{k+1}|_\chi - \sum_{j=0}^k |x|_{\tilde{\chi}_{x_j}} \geq \sup_x(|x|_\chi - \sum_{j=0}^k |x|_{\tilde{\chi}_{x_j}}) - \varepsilon_{k+1} > 0$ for some $\varepsilon_{k+1} < \varepsilon_0/2^{k+1}$.

Thus, either the iteration ends up in a finite number of steps and the result is proved, or the construction produces a sequence of points $\{x_j\}_{j=0}^\infty$, disjoint domains $\tilde{\Omega}_{x_j}$ with $\tilde{\Omega}_{x_j} = \Omega_{x_j} \setminus \cup_{i=0}^{j-1}\tilde{\Omega}_{x_i}$, and $\tilde{\chi}_{x_j} = \chi_{x_j}|_{\tilde{\Omega}_{x_j} \times \mathbb{R}^+}$. Let us prove that

$$|x|_\chi = \sum_{j=0}^\infty |x|_{\tilde{\chi}_{x_j}} \tag{7.9}$$

holds on Im χ. Otherwise there would be $x \in \mathrm{Im}\,\chi$ such that

$$|x|_\chi - \sum_{j=0}^\infty |x|_{\tilde{\chi}_{x_j}} > 0,$$

and therefore $\delta > 0$ such that for all k,

$$|\{\omega \notin \cup_{j=0}^k \tilde{\Omega}_{x_j} : x \in \mathrm{Im}\chi(\omega)\}| \geq |\{\omega \notin \cup_{j=0}^\infty \tilde{\Omega}_{x_j} : x \in \mathrm{Im}\chi(\omega)\}| \geq \delta > 0.$$

Since $\sum_{j=0}^\infty |\tilde{\Omega}_{x_j}| \leq |\Omega| < \infty$, $|\tilde{\Omega}_{x_j}| \to 0$ as $j \to \infty$. Now,

$$|\tilde{\Omega}_{x_{k+1}}| = |\{\omega \notin \cup_{j=0}^k \tilde{\Omega}_{x_j} : x_{k+1} \in \mathrm{Im}\chi(\omega)\}|$$

$$\geq \sup_x |\{\omega \notin \cup_{j=0}^k \tilde{\Omega}_{x_j} : x \in \mathrm{Im}\chi(\omega)\}| - \varepsilon_{k+1} \geq \delta - \varepsilon_{k+1} \geq \frac{\delta}{2}$$

for k large enough which yields a contradiction. Thus,

$$E^\alpha(\chi_{|\Omega\setminus(\cup_{j=0}^\infty \tilde{\Omega}_{x_j})}) = \int_{\Omega\setminus(\cup_{j=0}^\infty \tilde{\Omega}_{x_j})} \int_0^\infty |\chi(\omega,t)|^{\alpha-1}_{\chi_{|\Omega\setminus(\cup_{j=0}^\infty \tilde{\Omega}_{x_j})}}\, d\omega dt$$

$$< E^\alpha(\chi) < \infty.$$

By (7.9) we have $|\chi(\omega,t)|_{\chi_{|\Omega\setminus(\cup_{j=0}^\infty \tilde{\Omega}_{x_j})}} = 0$. All fibers having positive length we deduce that $|\Omega \setminus (\cup_{j=0}^\infty \tilde{\Omega}_{x_j})| = 0$.

The next proposition roughly states that optimal traffic plans can be monotonically approximated by finite irrigation graphs. It has also been proved in [58].

Proposition 7.14. *Let $\alpha \in [0,1)$ and $\chi : \Omega \to \mathbb{R}^+$ an optimal traffic plan. Assume that χ is parameterized by arc-length. Then there is a sequence ξ_n of restrictions of χ such that:*

- *each ξ_n is a finite irrigation graph;*
- *$\mathcal{D}(\xi_n) \subset \mathcal{D}(\xi_{n+1})$, $T_{\xi_n}(\omega) \to T_\chi(\omega)$ a.e.;*
- *$E^\alpha(\xi_n) \to E^\alpha(\chi)$;*
- *$\mu^+(\xi_n) \to \mu^+(\chi)$, $\mu^-(\xi_n) \to \mu^-(\chi)$.*

Finally, by changing Ω in a zero measure set, the restriction ξ of χ to the domain $\{(\omega, t), 0 < t < T(\omega)\}$ is strictly single path. In consequence, for every path $\chi(\omega, t)$ and every $0 < t_0 \leq t_1 < T(\omega)$ one has

$$|\Omega_{\overrightarrow{\chi(\omega,t_0),\chi(\omega,t_1)}}| > 0$$

and therefore $\inf_{t \in [t_0, t_1]} |\chi(\omega, t)|_\chi > 0$.

Proof. By Lemma 7.13, Ω is a disjoint union of Ω_j. Each restriction $\chi^j = \chi_{\Omega_j}$ is loop-free and single path by Proposition 7.4. Moreover, by substracting a set of zero measure from Ω_j, the restriction ξ^j of χ^j to the domain $\{(\omega, t) : \omega \in \Omega_j, 0 < t < T(\omega)\}$ is strictly single path. Each χ^j satisfies the assumptions of Proposition 7.11 and is in particular approximated by finite trees ξ_n^j. Thus one can build a monotone approximating sequence of finite graphs to χ by setting $\xi_n = \cup_{j=1}^n \xi_n^j$. All announced convergence properties follow by monotone convergence except one, the strictly single path property. Let us enforce it. This property is true for each ξ^j. Consider two finite trees ξ^j and ξ^k. Let ω_j and ω_k be two fibers in these trees which meet at two points x and y. Since ξ^j and ξ^k are strictly single path, both Ω_{xy}^j and Ω_{xy}^k have positive multiplicity. Since χ is single path they coincide between x and y. Since $\{(\omega, t) : \omega \in \Omega, 0 < t < T(\omega)\} = \cup_j \{(\omega, t) : \omega \in \Omega_j, 0 < t < T(\omega)\}$, we conclude that ξ is strictly single path.

Definition 7.15. *We say that a traffic plan χ is normal if it is loop-free, strictly single path and if for any fiber ω, $|\chi(\omega, t)|_\chi$ is bounded away from zero on any compact set contained in $]0, T(\omega)[$.*

According to Proposition 7.14, one can transform a traffic plan into a normal traffic plan by modifying its domain on a set with zero measure. Observe that in this case we have $\{(\omega, t) : \omega \in \Omega, 0 < t < T(\omega)\} \subset S_\chi$. Thus in the sequel all optimal traffic plans we consider will be assumed normal.

7.4 Bi-Lipschitz Regularity

The proof of the bi-Lipschitz regularity of fibers with positive flow follows [95].

Proposition 7.16. *Let χ be an optimal, normal traffic plan and x_0, x_1 two points such that $m := |\Omega_{\overrightarrow{x_0 x_1}}| > 0$. Denote by $\Gamma_{x_0 x_1}$ the support of the paths joining x_0 to x_1. Then*

$$\mathcal{H}^1(\Gamma_{x_0 x_1}) \left(\left(\frac{1}{m}\right)^\alpha - \left(\frac{1}{m} - 1\right)^\alpha \right) \leq ||x_0 - x_1||.$$

In consequence for every path $\chi(\omega, t)$ and every $0 < t_0 \leq t_1 < T_\chi(\omega)$,

$$c|t_1 - t_0| \leq ||\chi(\omega, t_0) - \chi(\omega, t_1)|| \leq |t_1 - t_0|$$

meaning that every path $t \mapsto \chi(\omega, t)$ is bi-Lipschitz with upper constant 1 and lower constant $c = \left(\frac{1}{m}\right)^\alpha - \left(\frac{1}{m} - 1\right)^\alpha$ where $m = |\Omega_{\overrightarrow{\chi(\omega,t_0)\chi(\omega,t_1)}}|$.

Proof. Consider an alternative traffic plan $\tilde{\chi}$ to χ which makes a straight shortcut between x_0 and x_1. This means that for every $\omega \in \Omega_{\overrightarrow{x_0 x_1}}$ we consider $t_0 < t_1$ such that $\chi(\omega, t_0) = x_0$ and $\chi(\omega, t_1) = x_1$ and set

$$\tilde{\chi}(\omega, t) = \begin{cases} \frac{t_1 - t}{t_1 - t_0} x_0 + \frac{t - t_0}{t_1 - t_0} x_1 & \text{if } \omega \in \Omega_{\overrightarrow{x_0 x_1}} \text{ and } t \in [t_0, t_1]; \\ \chi(\omega, t) & \text{otherwise.} \end{cases}$$

Then by the subadditivity of $s \mapsto s^\alpha$ and since χ is optimal,

$$0 \le E^\alpha(\tilde{\chi}) - E^\alpha(\chi) \le \int_{\Gamma_{x_0 x_1}} \left((|x|_\chi - m)^\alpha - |x|_\chi^\alpha \right) d\mathcal{H}^1(x) + ||x_0 - x_1|| m^\alpha,$$

where we write $m := |\Omega_{\overrightarrow{\chi(\omega, t_0)\chi(\omega, t_1)}}|$. This yields

$$\int_{\Gamma_{x_0 x_1}} \left(\left(\frac{|x|_\chi}{m} \right)^\alpha - \left(\frac{|x|_\chi}{m} - 1 \right)^\alpha \right) d\mathcal{H}^1(x) \le ||x_0 - x_1||.$$

Using the fact that the function $s \mapsto s^\alpha - (s - 1)^\alpha$ is decreasing on $[s, +\infty]$ we deduce that

$$\mathcal{H}^1(\Gamma_{x_0 x_1}) \left(\left(\frac{1}{m} \right)^\alpha - \left(\frac{1}{m} - 1 \right)^\alpha \right) \le ||x_0 - x_1||.$$

8 Interior and Boundary Regularity

In this chapter we first define connected components of an optimal traffic plan χ (Section 8.1), and the sub-traffic plans obtained by cutting χ at a point $x \in S_\chi$ (Section 8.2). These definitions and the fact that connected components of an optimal traffic plan are themselves optimal traffic plans will permit to perform some surgery leading to the main regularity theorems. The first "interior" regularity theorem (Section 8.3) states that outside the supports of μ^+ and μ^- an optimal traffic plan for the irrigation problem is a locally finite graph. This result is not proved in full generality, but under the restriction that either μ^+ or μ^- is finite atomic, or their supports are disjoint. What happens inside the support of the measures? Several properties can be established. Section 8.4 proves that at each branching point the number of branches is uniformly bounded by a constant depending only on α and the dimension N. The terms *interior and boundary regularity* for irrigation networks were proposed by Xia [96], [95]. Some techniques are borrowed from these papers. Yet, most of the statements and proofs follow [13].

8.1 Connected Components of a Traffic Plan

Definition 8.1. *Let χ be an optimal and normal traffic plan. Given an open set $V \subset \mathbb{R}^N$, two points x, y are said to be F-connected in V if there is a chain $x_1 = x, ..., x_l = y$ such that $\Gamma_{x_i x_{i+1}} \subset V$.*

The F-connection of two points in V defines an equivalence relation. The F-connected components[1] of χ in V are the classes obtained with this equivalence relation in $\operatorname{Im}(\chi) \cap V$ (see Figure 8.1). They are pairwise disjoint. When $V = \mathbb{R}^N$ we simply talk of F-connected components of χ.

Remark 8.2. Let χ be an optimal and normal traffic plan for the irrigation problem. Then each connected component T_i of χ in V can be parameterized as a restriction of χ. Indeed, let Ω_i be the set of fibers ω such that $\chi(\omega, \cdot)$ intersects T_i. Let $]s_i(\omega), t_i(\omega)[$ its maximal interval contained in V. Since the optimal traffic plan χ has no circuit, each fiber meets T_i at most once,

---

[1] The notion of F-connected component is obviously more restrictive than the notion of arc-connected component and less restrictive than the notion of indecomposable current.

M. Bernot et al., *Optimal Transportation Networks*. Lecture Notes in Mathematics 1955.

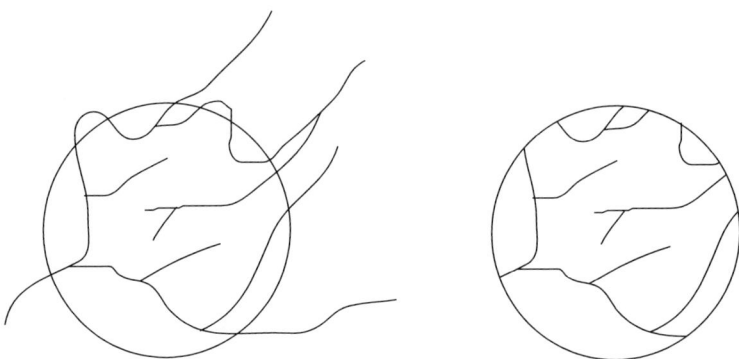

Fig. 8.1. A traffic plan χ is represented on the left-hand side. Following Definition 8.1, the figure on the right-hand side shows the F-connected components of χ in the ball $B(x, r)$. Notice that even if χ is globally F-connected, there are four F-connected components of χ in $B(x, r)$. The same fiber can contribute to several F-connected components. Yet if χ is optimal for the irrigation problem it has no circuit and each F-connected component contains at most one interval of each fiber. Thus each F-connected component of χ is a restriction of χ.

i.e. $\chi(\omega, \mathbb{R}^+ \backslash]s_i(\omega), t_i(\omega)[)$ does not intersect T_i. Thus we can parameterize T_i as a restriction of χ to $\cup_{\omega \in \Omega_i} \{\omega\} \times]s_i(\omega), t_i(\omega)[$, and we can define as χ_V the restriction of χ to the union of all these sets.

Definition 8.3. *We say that a family of traffic plans* $\{\chi_i\}_{i \in I}$ *is disjoint if* $E^\alpha(\cup_{i \in I} \chi_i) = \sum_{i \in I} E^\alpha(\chi_i)$.

Of course if the supports of a family of traffic plans (χ_i) are pairwise disjoint then the family is disjoint in the sense of the above definition. It is easily seen by the subadditivity of $s \mapsto s^\alpha$ that if a family of traffic plans is disjoint any subfamily also is disjoint.

Lemma 8.4. *Let* χ *be an optimal traffic plan for the irrigation problem and* V *be an open set in* $\mathbb{R}^N$. *The* F-connected components $\{\chi_i\}$ *of* χ *in* V *are optimal traffic plans that satisfy* $E^\alpha(\cup_{i \in I} \chi_i) = \sum_{i \in I} E^\alpha(\chi_i)$, *i.e. they are disjoint. Any finite union of* F-connected components is also optimal.

Proof. By Remark 8.2, each F-connected component of χ can be viewed as a traffic plan χ_i. Let us write $|x|_i := |x|_{\chi_i}$. Observe that $|x|_i > 0$ implies that $|x|_j = 0$ for all $j \neq i$. Indeed if we had $|x|_j > 0$ when $j \neq i$, then $\mathrm{Im}\,\chi_i$ and $\mathrm{Im}\,\chi_j$ would intersect and then they would coincide. This implies that $|x|_\chi = \sum_i |x|_i$. Thus,

$$E^\alpha(\chi_V) = \int_{V \cap \mathrm{Im}\,\chi} |x|_\chi^\alpha \, d\mathcal{H}^1 = \int_{V \cap \mathrm{Im}\,\chi} \sum_i |x|_i^\alpha \, d\mathcal{H}^1 = \sum_i E^\alpha(\chi_i).$$

Let us prove that all F-connected components χ_i are optimal. Assume by contradiction that one of them, χ_i, is not optimal. Then there is

$\xi_i \in TP(\mu^+(\chi_i), \mu^-(\chi_i))$ such that $E^\alpha(\xi_i) < E^\alpha(\chi_i)$. Consider the restrictions χ_i^1 and χ_i^2 of χ to $\cup_{\omega \in \Omega_i} \{\omega\} \times]0, s_i(\omega)[$ and $\cup_{\omega \in \Omega_i} \{\omega\} \times]t_i(\omega), +\infty[$ respectively. Then $\mu^-(\chi_i^1) = \mu^+(\chi_i) = \mu^+(\xi_i)$ and $\mu^+(\chi_i^2) = \mu^-(\chi_i) = \mu^-(\xi_i)$. Thus we can define by Lemma 5.5 the concatenation of χ_i^1, ξ_i and χ_i^2 which we denote by $\tilde{\chi}_i$. It is a traffic plan defined on Ω_i. Finally define a competitor $\tilde{\chi}$ to χ by setting $\tilde{\chi}(\omega) = \tilde{\chi}_i(\omega)$ if $\omega \in \Omega_i$ and $\tilde{\chi}(\omega) = \chi(\omega)$ otherwise. Notice that

$$E^\alpha(\chi) = \int_{\mathbb{R}^N} (|x|_{\chi|\Omega_i^c} + |x|_{\chi_i^1} + |x|_{\chi_i^2} + |x|_{\chi_i})^\alpha d\mathcal{H}^1$$

$$= \int_{\mathbb{R}^N} \left[(|x|_{\chi|\Omega_i^c} + |x|_{\chi_i^1} + |x|_{\chi_i^2})^\alpha + |x|_{\chi_i}^\alpha \right] d\mathcal{H}^1,$$

because the supports of χ_i on the one side and $\chi_{|\Omega_i^c}$, χ_i^1 and χ_i^2 on the other are disjoint by construction. By this last relation and subadditivity of $s \mapsto s^\alpha$,

$$E^\alpha(\tilde{\chi}) = \int_{\mathbb{R}^N} (|x|_{\chi|\Omega_i^c} + |x|_{\chi_i^1} + |x|_{\chi_i^2} + |x|_{\xi_i})^\alpha d\mathcal{H}^1$$

$$\leq E^\alpha(\chi) - E^\alpha(\chi_i) + E^\alpha(\xi_i)$$

$$< E^\alpha(\chi),$$

which yields a contradiction to the optimality of χ. The above proof is easily extended to prove the optimality of a finite union of k F-connected components.

8.2 Cuts and Branching Points of a Traffic Plan

Definition 8.5. *Assume that χ is an optimal, normal, and F-connected traffic plan. We call cuts of χ at x the F-connected components of χ in $\mathbb{R}^N \setminus \{x\}$.*

Observe that if $x \notin S_\chi$ then χ is the only cut at x. Observe also that, by Lemma 8.4, the cuts of χ at x are optimal and disjoint traffic plans and, having assumed that χ is normal, there can be countably many of them.

Proposition 8.6. *Let χ be optimal, normal and connected. Given a cut χ_i at $x \in S_\chi$, either $\mu^+(\chi_i)$ or $\mu^-(\chi_i)$, but not both, has an atomic mass at x. We call χ_i after-cut or before-cut accordingly (see Figure 8.2).*

Proof. Let χ_i be a cut at $x \in S_\chi$. If both $\mu^+(\chi_i)$ and $\mu^-(\chi_i)$ have an atomic mass at x, then χ would contain a circuit passing by x and therefore would not be optimal. If none of $\mu^+(\chi_i)$ and $\mu^-(\chi_i)$ have an atomic mass at x, then χ_i is a connected component of χ and therefore equal to χ and this is not possible if $x \in S_\chi$.

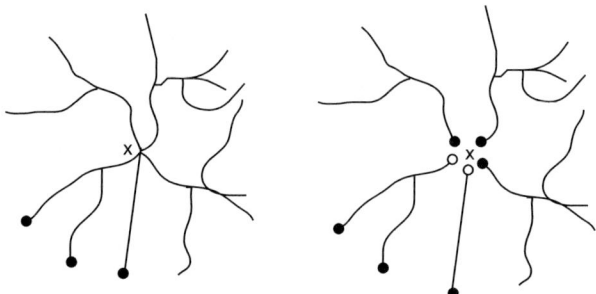

Fig. 8.2. Following Definition 8.5, the F-connected components of a traffic plan χ in $\mathbb{R}^N \setminus \{x\}$ define the "cuts" at a point x. These components can be separated in two classes depending on whether their mass at x is irrigating or irrigated (black or white disks on the figure).

Definition 8.7. Let χ be an optimal traffic plan. If χ is F-connected, a branching point of χ is a point $x \in S_\chi$ such that there are at least two cuts of χ at x of the same kind (after-cut or before-cut). We call branches of χ at x the before-cuts or after-cuts at x. Observe that if x is a branching point such that neither μ^+ nor μ^- contain a Dirac mass at x, then χ has at least three cuts. If χ is not F-connected we define the branching points of χ as the branching points of its F-connected components.

8.3 Interior Regularity

8.3.1 The Main Lemma

Lemma 8.8. Let (χ_n) be a family of disjoint traffic plans such that $\sum_n |\chi_n| < \infty$ and $\chi = \cup_n \chi_n$ is optimal. Then for any $J \subset \mathbb{N}$ the sub-traffic plan $\cup_{j \in J} \chi_j$ is also optimal.

Proof. Set $\mathbb{N} = I \cup J$, a disjoint union and $\chi_I = \cup_{i \in I} \chi_i$, $\chi_J = \cup_{j \in J} \chi_j$. By the disjointness of $(\chi_n)_n$ and subadditivity of $s \mapsto s^\alpha$,

$$E^\alpha(\chi_I) + E^\alpha(\chi_J) \leq \sum_{i \in I} E^\alpha(\chi_i) + \sum_{j \in J} E^\alpha(\chi_j) = E^\alpha(\chi) = E^\alpha(\chi_I \cup \chi_J)$$
$$\leq E^\alpha(\chi_I) + E^\alpha(\chi_J).$$

Thus

$$E^\alpha(\chi_I) + E^\alpha(\chi_J) = E^\alpha(\chi). \tag{8.1}$$

Let $\tilde{\chi}_J$ be a competitor to χ_J with the same irrigated measures. Then $\tilde{\chi} = \tilde{\chi}_J \cup \chi_I$ has the same irrigated measure as χ. Since χ is optimal, using (8.1) and again the subadditivity,

$$E^\alpha(\chi) = E^\alpha(\chi_I) + E^\alpha(\chi_J) \le E^\alpha(\tilde{\chi}) \le E^\alpha(\chi_I) + E^\alpha(\tilde{\chi}_J).$$

Thus $E^\alpha(\chi_J) \le E^\alpha(\tilde{\chi}_J)$ which proves χ_J to be optimal.

Lemma 8.9. *Let $\alpha \in (1 - \frac{1}{N}, 1)$ and $(\chi_n)_{n=1}^\infty$ a sequence of disjoint traffic plans such that $\sum_n |\chi_n| < \infty$ and*

$$\mathrm{dist}(\overline{\cup_n \mathrm{supp}(\mu^+(\chi_n))}, \overline{\cup_n \mathrm{supp}(\mu^-(\chi_n))}) > 0.$$

Then there is some n or some pair (n, m) such that χ_n or $\chi_n \cup \chi_m$ is not optimal. Therefore, $\cup_n \chi_n$ is not optimal.

Proof. If some χ_n is not optimal there is nothing to prove. Thus assume in the sequel that all χ_n are optimal and that $|\chi_n|$ is decreasing. Consider the renormalized sequence $\tilde{\chi}_n \in \mathrm{TP}(\mu^+(\tilde{\chi}_n), \mu^-(\tilde{\chi}_n))$ so that $\mu^+(\tilde{\chi}_n) := \frac{\mu^+(\chi_n)}{|\chi_n|}$ and $\mu^-(\tilde{\chi}_n) := \frac{\mu^-(\chi_n)}{|\chi_n|}$. By extraction of a subsequence using Theorem 3.28, we may assume that $\tilde{\chi}_n$ converges to $\tilde{\chi}$. Notice that $\mu^+(\tilde{\chi}_n)$ and $\mu^-(\tilde{\chi}_n)$ weakly converge to the probability measures $\tilde{\mu}^+ = \mu^+(\tilde{\chi})$ and $\tilde{\mu}^- = \mu^-(\tilde{\chi})$ respectively. Observe that $\mathrm{supp}(\tilde{\mu}^+) \subseteq \overline{\cup_n \mathrm{supp}(\tilde{\mu}^+(\chi_n))}$ and $\mathrm{supp}(\tilde{\mu}^-) \subseteq \overline{\cup_n \mathrm{supp}(\tilde{\mu}^-(\chi_n))}$, hence

$$\mathrm{dist}(\mathrm{supp}(\tilde{\mu}^+), \mathrm{supp}(\tilde{\mu}^-)) > 0. \tag{8.2}$$

The stability Proposition 6.12 states that the limit of a converging sequence of optimal traffic plans is optimal. In particular, $E^\alpha(\tilde{\chi}_n)$ is a Cauchy sequence, that is

$$E^\alpha(\tilde{\chi}_n) - E^\alpha(\tilde{\chi}_m) = \varepsilon_0(n, m), \tag{8.3}$$

where $\varepsilon_0(n, m) \to 0$ when $m, n \to \infty$. Moreover, $E^\alpha(\tilde{\chi}_n) \to E^\alpha(\tilde{\chi})$ where $E^\alpha(\tilde{\chi}) > 0$ (otherwise $E^\alpha(\tilde{\mu}^+, \tilde{\mu}^-) = 0$, and this contradicts (8.2)). Returning to χ_m and χ_n (8.3) implies that

$$\left| \frac{E^\alpha(\chi_m)}{E^\alpha(\chi_n)} - \frac{|\chi_m|^\alpha}{|\chi_n|^\alpha} \right| = \varepsilon_1(m, n) \frac{|\chi_m|^\alpha}{|\chi_n|^\alpha},$$

where $\varepsilon_1(n, m) \to 0$ when $m, n \to \infty$. In addition, $E^\alpha(\mu^+(\tilde{\chi}_m), \mu^+(\tilde{\chi}_n))$ and $E^\alpha(\mu^-(\tilde{\chi}_m), \mu^-(\tilde{\chi}_n))$ tend to zero as $m, n \to \infty$, so that

$$E^\alpha(\mu^-(\chi_m), \frac{|\chi_m|}{|\chi_n|} \mu^-(\chi_n)) = \varepsilon_2(n, m) |\chi_m|^\alpha,$$

and

$$E^\alpha(\mu^+(\chi_m), \frac{|\chi_m|}{|\chi_n|} \mu^+(\chi_n)) = \varepsilon_3(n, m) |\chi_m|^\alpha,$$

where $\varepsilon_2(n, m), \varepsilon_3(n, m) \to 0$ when $m, n \to \infty$. As a last preliminary estimate we note that for n large enough $E^\alpha(\tilde{\chi}_n) \ge c := \frac{E^\alpha(\tilde{\chi})}{2}$. Thus returning to χ_n,

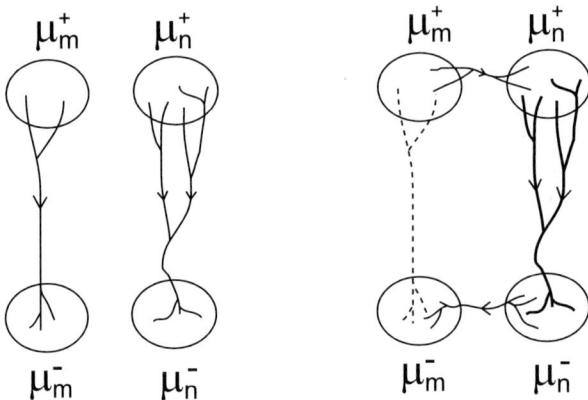

Fig. 8.3. Illustrates the proof of Lemma 8.9. Under the assumptions of Lemma 8.9, there is a shortcut χ_m through χ_n that has a better cost than the union of χ_m with χ_n that is represented on the left-hand side. The traffic plan represented on the right-hand side is the shortcut of χ_m through χ_n, i.e. all the mass that was transported by χ_m is transported through χ_n.

$$E^\alpha(\chi_n) \geq c|\chi_n|^\alpha. \tag{8.4}$$

We shall prove that some $\chi_m \cup \chi_n$ is not optimal by constructing a short-cut between χ_n and χ_m (see Figure 8.3). In the shortcut the mass that was transported by χ_m shall be transported through a rescaled χ_n. This is done by first transporting the measure $\mu^+(\chi_m)$ to $\frac{|\chi_m|}{|\chi_n|}\mu^+(\chi_n)$ with a cost $E^\alpha(\mu^+(\chi_m), \frac{|\chi_m|}{|\chi_n|}\mu^+(\chi_n)) = \varepsilon_3(n,m)|\chi_m|^\alpha$; second the traffic plan χ_n has its mass dilated by the factor $\frac{|\chi_m|+|\chi_n|}{|\chi_n|}$; third the measure $\frac{|\chi_m|}{|\chi_n|}\mu^-(\chi_n)$ is trans-ported to $\mu^-(\chi_m)$ with a cost $E^\alpha(\mu^-(\chi_m), \frac{|\chi_m|}{|\chi_n|}\mu^-(\chi_n)) = \varepsilon_2(n,m)|\chi_m|^\alpha$. This concatenation defines by Lemma 5.5 a traffic plan from $\mu^+(\chi_m)$ to $\mu^-(\chi_m)$ which we call shortcut of χ_m through χ_n and denote by $\mathrm{sc}(\chi_m, \chi_n)$. Its cost is

$$E^\alpha(\mathrm{sc}(\chi_m, \chi_n)) = (\varepsilon_2(n,m) + \varepsilon_3(n,m))|\chi_m|^\alpha + (1 + \frac{|\chi_m|}{|\chi_n|})^\alpha E^\alpha(\chi_n)$$

$$\leq (\varepsilon_2(n,m) + \varepsilon_3(n,m))|\chi_m|^\alpha + (1 + \alpha\frac{|\chi_m|}{|\chi_n|})E^\alpha(\chi_n).$$

On the other hand, since all the traffic plans χ_n are disjoint one gets by using (8.3)

$$E^\alpha(\chi_n \cup \chi_m) = E^\alpha(\chi_n) + E^\alpha(\chi_m) \geq E^\alpha(\chi_n)\left(1 + (1 - \varepsilon_1(m,n))\frac{|\chi_m|^\alpha}{|\chi_n|^\alpha}\right),$$

so that using (8.4),

$$\delta E^\alpha := E^\alpha(\mathrm{sc}(\chi_n, \chi_m)) - E^\alpha(\chi_n \cup \chi_m)$$
$$\leq (\varepsilon_2(n,m) + \varepsilon_3(n,m))|\chi_m|^\alpha$$
$$+ (\alpha\frac{|\chi_m|}{|\chi_n|} - (1 - \varepsilon_1(m,n))\frac{|\chi_m|^\alpha}{|\chi_n|^\alpha})E^\alpha(\chi_n)$$
$$\leq \left[\alpha\frac{|\chi_m|}{|\chi_n|} - (1 - \varepsilon_1 - c^{-1}(\varepsilon_2 + \varepsilon_3)))\left(\frac{|\chi_m|}{|\chi_n|}\right)^\alpha\right]E^\alpha(\chi_n)$$
$$< 0.$$

The last inequality is obtained by taking m and n large enough to have $(1 - \varepsilon_1 - c^{-1}(\varepsilon_2 + \varepsilon_3)) > \frac{1}{2}$ and then fixing n and growing m to ensure that $\frac{|\chi_m|}{|\chi_n|}$ is small enough. This proves that $\chi_m \cup \chi_n$ cannot be optimal. Thus $\cup_n \chi_n$ is not optimal by Lemma 8.8.

8.3.2 Interior Regularity when $\overline{\mathrm{supp}(\mu^+)} \cap \overline{\mathrm{supp}(\mu^-)} = \emptyset$

Lemma 8.10. *Let $\alpha \in (1 - \frac{1}{N}, 1)$ and let χ be an optimal traffic plan in $\mathrm{TP}(\mu^+, \mu^-)$ such that $E^\alpha(\chi) < \infty$. Assume that the supports of μ^+ and μ^- are at positive distance. Then χ has a finite number of F-connected components.*

Proof. If χ has an infinite number of F-connected components χ_n, Lemma 8.4 states that the χ_n are disjoint traffic plans such that $\chi = \cup_n \chi_n$. In addition,

$$\mathrm{dist}(\overline{\cup_n \mathrm{supp}(\mu^+(\chi_n))}, \overline{\cup_n \mathrm{supp}(\mu^-(\chi_n))}) \geq \mathrm{dist}(\overline{\mathrm{supp}(\mu^+)}, \overline{\mathrm{supp}(\mu^-)})$$
$$> 0,$$

so that the optimality of χ contradicts Lemma 8.9.

Lemma 8.11. *Let $\alpha \in (1 - \frac{1}{N}, 1)$ and let χ be an optimal traffic plan in $\mathrm{TP}(\mu^+, \mu^-)$ such that $E^\alpha(\chi) < \infty$. Assume that the supports of μ^+ and μ^- are at positive distance. Then, for all $x \in \mathrm{Im}\,\chi$ such that $\mathrm{dist}(x, \mathrm{supp}(\mu^+) \cup \mathrm{supp}(\mu^-)) > 0$ the number of cuts of χ at x is finite.*

Proof. The result of the present lemma could have been derived from the stronger Theorem 8.20 in the next section. We give here a direct and simple argument. Let us assume that χ has an infinite number of cuts χ_x^n at x. Without restriction we can consider that all of the cuts χ_x^n are cuts after x so that by definition of χ_x^n,

$$\mathrm{dist}(\overline{\cup_n \mathrm{supp}(\mu^+(\chi_x^n))}, \overline{\cup_n \mathrm{supp}(\mu^-(\chi_x^n))}) \geq \mathrm{dist}(\overline{\mathrm{supp}(\mu^+) \cup \{x\}}, \overline{\mathrm{supp}(\mu^-)})$$
$$> 0.$$

In addition, Lemma 8.4 applied to $V = \mathbb{R}^N \setminus \{x\}$ and χ states that the χ_x^n are disjoint traffic plans such that every finite union of them is optimal. This contradicts Lemma 8.9.

Lemma 8.12. *Let $\alpha \in (1 - \frac{1}{N}, 1)$ and let χ be an optimal and normal traffic plan in $\mathrm{TP}(\mu^+, \mu^-)$ such that $E^\alpha(\chi) < \infty$. Assume that the supports of μ^+ and μ^- are at positive distance. Then there are finitely many branching points in any ball $B(x, r)$ outside of the support of μ^+ and μ^-.*

Proof. By Lemma 8.10, χ has a finite number of F-connected components. Thus, it is sufficient to prove the lemma for a F-connected traffic plan χ. Let us suppose that there is a sequence of distinct branching points $x_n \in B(x, r)$ such that $x_n \to x$ and $B(x, r)$ does not meet the support of μ^- and μ^+. We shall prove that we can construct from χ an infinite disjoint sequence of traffic plans that are cuts at points in the sequence x_n.

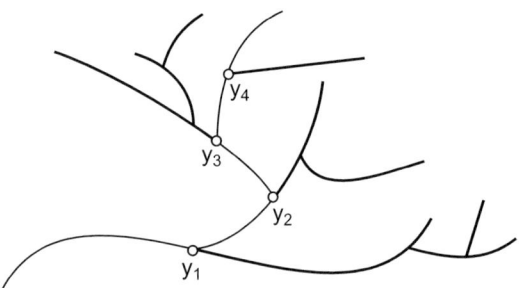

Fig. 8.4. Proof of Lemma 8.12. If there are infinitely many branching points x_i, one can find infinitely many disjoint cuts at some subsequence y_i. On this figure, the infinite sequence of disjoint traffic plans $\chi^2_{y_i}$ is represented by thicker lines.

Let us first consider the branching point $y_1 = x_1$. By definition, there are, at least, three cuts of χ at y_1 (by Lemma 8.11 there is a finite number of them); one of the cuts has necessarily an infinite number of branching points in $B(x, r)$. Denote it by $\chi^1_{y_1}$ and denote by $\chi^2_{y_1}$ another cut. Let us consider a fiber $\chi^1_{y_1}(\omega_1, \cdot)$. If it contains infinitely many branching points x_{i_j} in $B(x, r)$, then one may choose at any of them a cut. This yields infinitely many cuts of χ at points in the sequence $\{x_n\}$. These cuts are disjoint, since there are no circuits in χ. If now the fiber $\chi^1_{y_1}(\omega_1, \cdot)$ contains a finite number of branching points in $B(x, r)$, take the first (starting from y_1), and let us call this point y_2 (see Figure 8.4).

Again, $\chi^1_{y_1}$ has at least three cuts at y_2 (and at most a finite number): one of them contains y_1 and one of the other cuts, call it $\chi^1_{y_2}$, contains infinitely many branching points in $B(x, r)$. Select a third one and denote it by $\chi^2_{y_2}$. Again consider some fiber $\chi^1_{y_2}(\omega_1, \cdot)$. If the fiber contains infinitely many branching points in the sequence $\{x_n\}$ we are done, otherwise we proceed as in the previous paragraph. By iterating the previous argument, if we never find a fiber that contains infinitely many branching points in the sequence $\{x_n\}$, then we construct a disjoint sequence of cuts $\chi^2_{y_i}$ at points y_i in the sequence $\{x_n\}$.

By extracting a subsequence, if necessary, we may assume that all the cuts $\chi^2_{y_i}$ are after-cuts. Moreover, since y_i converges to a point x far from $\text{supp}(\mu^+(\chi)) \cup \text{supp}(\mu^-(\chi))$ and

$$\text{dist}(\text{supp}(\mu^+(\chi)), \text{supp}(\mu^-(\chi))) > 0,$$

we have

$$\text{dist}(\overline{\cup_i \text{supp}(\mu^+(\chi^2_{y_i}))}, \overline{\cup_i \text{supp}(\mu^-(\chi^2_{y_i}))}) > 0.$$

Thus we can apply Lemma 8.9 and deduce that there exists $\chi^2_{y_m}$ and $\chi^2_{y_n}$ such that the union $\chi^2_{y_m} \cup \chi^2_{y_n}$ is not optimal. Now these traffic plans are F-connected components of χ in $\mathbb{R}^N \setminus \{y_m, y_n\}$. Thus χ is not optimal by Lemma 8.4, a contradiction.

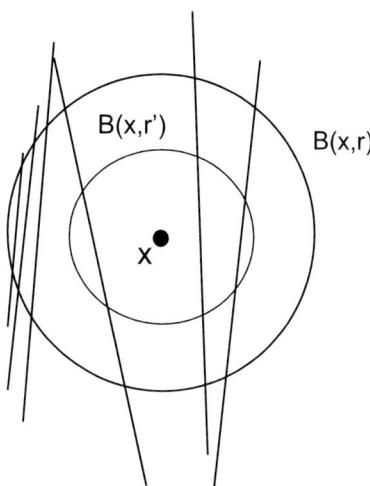

Fig. 8.5. Proof of Lemma 8.13. This figure represents the balls $B(x,r)$ and $B(x, r\cos(\arcsin(\frac{\varepsilon}{2r})))$. We observe that the segments in $B(x,r)$ with length lower than ε do not cross the ball $B(x, r\cos(\arcsin(\frac{\varepsilon}{2r})))$. Thus, if there are only finitely many segments with length larger than ε, there is only a finite number of segments crossing $B(x, r\cos(\arcsin(\frac{\varepsilon}{2r})))$.

Lemma 8.13. Let $\alpha \in (1 - \frac{1}{N}, 1)$ and let χ be an optimal traffic plan in $\text{TP}(\mu^+, \mu^-)$. Assume that the number of branching points of χ is finite in any closed ball $B(x,r)$ not meeting the supports of μ^+ and μ^-. Then the traffic plan has the structure of a finite graph in any of those balls.

Proof. For all $\omega \in \Omega$, $\chi(\omega, \cdot)^{-1}(B(x,r))$ can be decomposed into a disjoint union $\cup_i(t_i^+, t_i^-)$. By assumption either $\chi(\omega, (t_i^+, t_i^-))$ contains no branching point, or it has a finite number of them. In the case there is no branching point, the multiplicity at point $\chi(\omega, t)$ is constant for $t \in (t_i^+, t_i^-)$. Thus, by optimality of χ, $\chi(\omega, (t_i^+, t_i^-))$ is a segment. In the case $\chi(\omega, (t_i^+, t_i^-))$ has a finite number of branching points, the multiplicity along $\chi(\omega, t)$ for $t \in (t_i^+, t_i^-)$ is piecewise constant so that $\chi(\omega, (t_i^+, t_i^-))$ is piecewise linear.

Thus, the restriction of the support of χ to $B(x, r)$ is made of a finite graph structure and a possibly infinite number of disjoint segments connecting two points of $\partial B(x, r)$.

If we fix $\varepsilon > 0$, there is a finite number of segments with length larger than ε. Otherwise, by compactness of $\partial B(x, r)$, we would construct an infinite sequence of disjoint segment traffic plans χ_n such that $\mu^+(\chi_n)$ converges to δ_x and $\mu^-(\chi_n)$ converges to δ_y where $x, y \in \partial B(x, r)$ are such that $||x - y|| \geq \varepsilon$. This is ruled out by Lemma 8.9.

Thus, there are finitely many segments with length larger than ε. We now observe that in the ball $B(x, r \cos(\arcsin(\frac{\varepsilon}{2r})))$ all the segments (in $B(x, r)$) shorter than ε are eliminated (see Figure 8.5). Thus, χ has a finite graph structure made of segments in $B(x, r')$, where $r' = r \cos(\arcsin(\frac{\varepsilon}{2r}))$. We can apply the above conclusion to an open ball $B(x, \rho)$ with $\rho > r$ and $\rho - r$ small enough to ensure that $B(x, \rho)$ does not meet the supports of μ^+ and μ^-. Taking $\varepsilon < \rho - r$ finishes the proof.

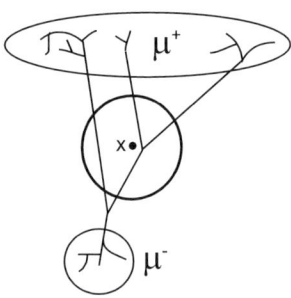

Fig. 8.6. Let $\alpha \in (1 - \frac{1}{N}, 1)$ and χ be an optimal traffic plan in $\mathrm{TP}(\mu^+, \mu^-)$, where the supports of μ^+ and μ^- are at positive distance. Theorem 8.14 states that χ has the structure of a finite graph in any ball $B(x, r)$ outside of the support of μ^+ and μ^-.

Theorem 8.14. *Let $\alpha \in (1 - \frac{1}{N}, 1)$ and let χ be an optimal traffic plan in $\mathrm{TP}(\mu^+, \mu^-)$. Assume that the supports of μ^+ and μ^- are at positive distance. In any closed ball $B(x, r)$ not meeting the supports of μ^+ and μ^-, the traffic plan has the structure of a finite graph (see Figure 8.6).*

Proof. This is a direct consequence of Lemmas 8.12 and 8.13.

8.3.3 Interior Regularity when μ^+ is a Finite Atomic Measure

Lemma 8.15. *Let* $\alpha \in (1 - \frac{1}{N}, 1)$ *and* $\chi \in TP(\delta_S, \mu^-(\chi))$, *where* S *is a source point, be an optimal traffic plan such that* $E^\alpha(\chi) < \infty$. *In any ball* $B(x, r)$ *outside the support of* $\mu^-(\chi)$, *the traffic plan* χ *has a finite number of branching points.*

Proof. Because of Proposition 7.4, we know that μ has the structure of a tree. By Definition 8.7, if we consider a branching point x with positive multiplicity, it induces (at least) two after-cuts that we denote by χ_x^1 and χ_x^2, i.e. cuts that do not contain the source point S.

Let us suppose by contradiction that there is a sequence of distinct branching points x_n such that $x_n \to x$ with x outside the support of $\mu^-(\chi)$. Let us prove that we can extract a subsequence of x_n along with disjoint cuts χ_{x_n} (see Figure 8.7). Consider two cases:

(i) Assume that there is an infinite subsequence $x_{\phi(n)}$ such that one of the subtrees $\chi_{x_{\phi(n)}}^1$ or $\chi_{x_{\phi(n)}}^2$ contains a finite number of x_j, say $\chi_{x_{\phi(n)}}^1$. Then select first $x_{\phi(1)}$ with the cut $\chi_{x_{\phi(1)}}^1$ and remove the finite number of points x_j contained in $\chi_{x_{\phi(1)}}^1$ from the sequence $(x_{\phi(n)})_n$. Proceed iteratively by selecting the next point $x_{\phi(j)}$ that is still in the list along with the cut $\chi_{x_{\phi(j)}}^1$ that has a finite number of branching points x_n. The list is never empty since we only remove a finite number of points from that list at each step.

(ii) Otherwise, all but a finite number of branching points x_i are such that $\chi_{x_i}^1$ and $\chi_{x_i}^2$ contain an infinite number of x_n. Call such branching points "good". Select a good x_1 along with $\chi_{x_1}^1$. Then select a good branching point in $\chi_{x_1}^2$ along with its χ^1 tree and proceed to further extraction.

In both cases we constructed a sequence of disjoint traffic plans χ_{x_i} such that $\mu^+(\chi_{x_i}) = m_i \delta_{x_i}$ and $\sum_i m_i \leq 1$. Since $x_i \to x$ with x outside the support of $\mu^-(\chi)$, x is outside $\cup_i \text{supp}(\mu^-(\chi_{x_i}))$ and for $i \geq i_0$ large enough the distance of $\cup_{i \geq i_0} \{x_i\}$ to $\cup_i \text{supp}(\mu^-(\chi_{x_i}))$ is positive. Applying Lemma 8.9 to the sequence $(\chi_{x_i})_{i \geq i_0}$ we deduce that there are n and m such that the union $\chi_{x_m} \cup \chi_{x_n}$ is not optimal. By Lemma 8.4 this yields a contradiction. Indeed, χ_{x_m} and χ_{x_n} are F-connected components of χ in the open set $\mathbb{R}^N \setminus \{x_m, x_n\}$. $\square$

Theorem 8.16. *Let* $\alpha \in (1 - \frac{1}{N}, 1)$ *and* χ *be an optimal traffic plan from* δ_S *to* $\mu^-(\chi)$. *In any ball* $B(x, r)$ *outside the support of* $\mu^-(\chi)$, *the traffic plan* χ *has a finite graph structure.*

Proof. This is a direct consequence of Lemmas 8.15 and 8.13. $\square$

Corollary 8.17. *Let* $\alpha \in (1 - \frac{1}{N}, 1)$ *and let* χ *be an optimal traffic plan such that* $\mu^+(\chi) = \sum_{i=1}^n m_i \delta_{x_i}$ *is an atomic measure. In any ball* $B(x, r)$ *outside the support of* $\mu^-(\chi)$, *the traffic plan* μ *has a finite graph structure.*

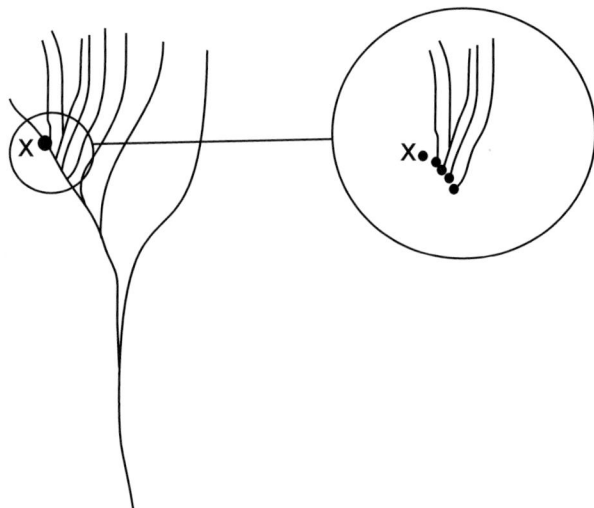

Fig. 8.7. Illustrates the proof of Lemma 8.15. If x is an accumulation point of branching points, one can cut from the tree an infinite sequence of disjoint subtrees. This is in contradiction with Lemma 8.9. The figure shows the particular case where all subtrees bifurcate from a single fiber.

Proof. We can proceed exactly as in Theorem 8.16. It is enough to prove the analogue of Lemma 8.15, namely that the traffic plan has a finite number of branching points in any ball outside the support of $\mu^-(\chi)$. Set $\Omega_i := \{\omega : \chi(\omega, 0) = x_i\}$ and $T_i := \chi(\Omega_i, \mathbb{R}^+)$. The T_i's are single trunk trees whose union is χ. By Proposition 7.8, T_i and T_j meet for $i \neq j$ at most on one fiber interval. Call x_{ij}, y_{ij} the extremities of these intervals. Then x_{ij} and y_{ij} are branching points of χ and their number cannot exceed $n(n-1)$. We call such points contact points. All other branching points of χ also are branching points of one of the trees T_i. Let us now suppose that there is an infinite number of branching points in some ball $B(x, r)$ outside the support of $\mu^-(\chi)$. Since there is only a finite number of contact points and a finite number of trees T_i, at least one of the trees has an infinite number of branching points. Thus the proof of Lemma 8.15 applies. One can cut from T_i an infinite number of disjoint traffic plans and obtain a contradiction with the optimality of χ.

The regularity of the irrigation problem was studied in [95, 96] where the main results were stated. The finite graph structure of the solution of the irrigation problem for atomic measures was also proved in [58]. The regularity of optimal transportation networks for urban planning has been studied in [22, 24].

8.4 Boundary Regularity

Let us fix $\alpha \in [0, 1]$. Let x, x_1, x_2 be such that $|x - x_1| = |x - x_2| = r$. Consider any irrigation network made of $[x, x_1]$ with flow m_1 and $[x, x_2]$ with flow m_2. (The flow can be directed from x to x_1 or from x_1 to x and the same for x_2.) We know from Proposition 4.5 in [96] (see also Lemma 12.2 or Lemma 6.6.2 in [10]) that if this network is optimal then the angle of $[x, x_1]$ with $[x, x_2]$ is larger than a minimal angle $\theta(\alpha, N)$. This angle does not depend upon r, m_1 or m_2. Let us call $\mathcal{N}(\alpha, N)$ the maximal number of points $x_1, \dots x_{\mathcal{N}(\alpha, N)}$ such that all segments $[x, x_i]$ make an angle larger than $\theta(\alpha, N)$. As a straightforward consequence of the above fact we can deduce the following lemma.

Lemma 8.18. *Let $(x_i)_{i=1}^l$ be distinct points on $\partial B(x, r)$ such that $l > \mathcal{N}(\alpha, N)$. Then there are x_j and x_k such that for every $m_j > 0$ and $m_k > 0$, the irrigation network made of the paths $[x, x_j]$ and $[x, x_k]$ with flows m_j and m_k is not optimal. In addition, the energy gain obtained by replacing this network by the optimal configuration from x to x_j and x_k is bounded from below by a function of the form $c(\alpha, N, m_j, m_k)r > 0$ where c is continuous with respect to (m_j, m_k).*

Proof. If $r = 1$ the energy gain is some constant $c(\alpha, N, m_j, m_k) > 0$. The result follows by a mere scaling.

Lemma 8.19. *Let χ an optimal traffic plan in $\mathrm{TP}(\delta_S, \mu^-)$ such that the cut of χ at S defines one F-connected component. Then, for all $\omega \in \Omega$, $|\chi(\omega, t)|_\chi \to 1$ when $t \to 0$.*

Proof. Since the cut of χ at S defines one F-connected component, each fiber ω' coincides with $\chi(\omega, t)$ on some interval $[0, t_{\omega'}^+(\omega)]$ with $t_{\omega'}^+(\omega) > 0$. Thus, $\Omega_{\chi(\omega,t)}$ is a decreasing sequence of sets with respect to t such that $\cup_{t>0} \Omega_{\chi(\omega,t)} = \Omega_x$. This implies that $|\chi(\omega, t)|_\chi = |\Omega_{\chi(\omega,t)}| \to 1$ when $t \to 0$.

The main result of this section is the following.

Theorem 8.20. (bounded branching property) *Let $\alpha \in (0, 1]$. At every point x of the support of an optimal traffic plan χ in $\mathbb{R}^N$, the number of branches at x is less than a constant $\mathcal{N}(\alpha, N)$ depending only on N and α.*

Proof. Let us first introduce some notations. Let $x \in S_\chi$ and consider the branches of χ at x (they are before-cuts or after-cuts) that we denote by χ^i, where $i \in I \subset \mathbb{N}$. Observe that the F-connected components of χ in the open set $B(x, r) \setminus \{x\}$ whose irrigation measure contains a Dirac mass at x are in one-to-one correspondence with the branches χ^i. We shall denote them by $\chi^{i,r}$. Let Ω_x^i be the set of fibers ω of χ^i such that $x \in \chi^i(\omega, \mathbb{R}^+)$. We shall denote by χ_x^i the restriction of χ^i to Ω_x^i. The multiplicity of χ_x^i at x is denoted by m_i (observe that $m_i = |\Omega_x^i|$). For each i and each r such that χ_x^i

exits the ball $B(x, r)$, there is a fiber ω such that the multiplicity of $\chi_x^i(\omega)$ at its first exit point is maximal among all $\omega \in \Omega_x^i$. We denote $m_i(r)$ this maximal multiplicity. Lemma 8.19 ensures that $m_i(r) \to m_i$ when $r \to 0$.

Assume by contradiction that the cardinality of I exceeds $\mathcal{N}(\alpha, N)$. Consider $r > 0$ small enough so that the first $\mathcal{N}(\alpha, N) + 1$ trees χ_x^i exit $B(x, r)$. Let $x_j(r)$ and $x_k(r)$ with multiplicities $m_j(r)$ and $m_k(r)$ such that $j < k \leq \mathcal{N}(\alpha, N) + 1$ and satisfying the conclusions of Lemma 8.18. The indices j and k depend upon r. By extracting a subsequence $r_n \to 0$ one can fix j and k. For a sake of simplicity we write in the sequel r for r_n or even omit r.

Consider the network χ^r made of the union of $\chi^{j,r}$ and $\chi^{k,r}$. By Lemma 8.4 such a traffic plan is defined on some $\Omega(r) \subset \Omega$ and is optimal. Consider two types of fibers in this traffic plan: the set $\Omega_1 \subset \Omega(r)$ of the fibers which connect x to x_j or x_k and $\Omega_2 = \Omega(r) \setminus \Omega_1$. Thus $|\Omega_1| = m_j(r) + m_k(r) \to m_j + m_k$ and $|\Omega_2| = \varepsilon(r)$ where $\varepsilon(r) \to 0$ when $r \to 0$.

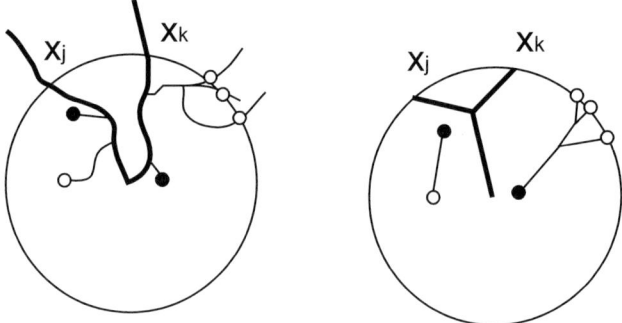

Fig. 8.8. Proof of Theorem 8.20. The main flows going through x_j and x_k are represented on the left hand side figure. The other part of the flow irrigates a measure $\nu(r)$ which is represented by black and white dots. The reconnection of x to x_j and x_k through an optimal configuration, along with an optimal transport of $\nu(r)$ prove that the number of cuts at any point of the support of a traffic plan is bounded by a number depending only on α and N.

Let us build a competitor $\tilde{\chi}^r$ to χ^r:

– Let $\nu(r)$ be the measure $\mu(\chi_{|\Omega_2}^r)$, which has support in $\overline{B(x, r)}$. Let χ_2 be an optimal network connecting $\nu(r)$ and define $\tilde{\chi}_{|\Omega_2}^r := \chi_2$.
– The network $\chi_{|\Omega_1}^r$ is the union of two paths between x and x_j, x_k with flows $m_j(r)$ and $m_k(r)$ respectively. Define $\tilde{\chi}_{|\Omega_1}^r$ as the optimal network between x, x_j and x_k with the same flows (see Figure 8.8).

Let us now see how the energy of the network is changed by this rearrangement. Since $\Omega(r) = \Omega_1 \cup \Omega_2$, by Lemma 5.2,

$$E^\alpha(\tilde{\chi}^r) \le E^\alpha(\tilde{\chi}^r_{|\Omega_1}) + E^\alpha(\tilde{\chi}^r_{|\Omega_2}). \tag{8.5}$$

– By Lemma 8.18 and since $m_j(r) \simeq m_j$ and $m_k(r) \simeq m_k$ and c is continuous, there is some $C > 0$ such that

$$E^\alpha(\tilde{\chi}^r_{|\Omega_1}) \le E^\alpha(\chi^r_{|\Omega_1}) - c(\alpha, N, m_j(r), m_k(r))r \le E^\alpha(\chi^r_{|\Omega_1}) - Cr. \tag{8.6}$$

– Since $|\Omega_2| = \varepsilon(r)$,

$$E^\alpha(\tilde{\chi}^r_{|\Omega_2}) \le C' r \varepsilon(r)^\alpha = o(r), \tag{8.7}$$

where C' denotes the maximal energy for irrigating any measure μ such that $|\mu^+| = |\mu^-| = 1$ in $B(0, 1)$.

From (8.5), (8.6), (8.7) and Lemma 5.2 we obtain

$$E^\alpha(\tilde{\chi}^r) \le E^\alpha(\chi^r_{|\Omega_1}) + o(r) - Cr < E^\alpha(\chi^r_{|\Omega_1}) \le E^\alpha(\chi^r)$$

for r small enough. Now, by Lemma 8.4 applied to $V = B(x, r) \setminus \{x\}$ and χ, χ_r is optimal. Since $\tilde{\chi}^r$ and χ^r irrigate the same measure the last inequality yields a contradiction.

8.4.1 Further Regularity Properties

In this short section we pile up two further structure properties for optimal traffic plans.

Corollary 8.21. *If χ is an optimal traffic plan such that $E^\alpha(\chi) < \infty$, Proposition 7.14 ensures that $S_\chi = \cup_j S_{\chi_j}$ where χ_j has a finite number of branching points. Thus, any optimal traffic plan χ such that $E^\alpha(\chi) < \infty$ has countably many branching points.*

Theorem 8.22. ("local irrigation property"). *If χ is a traffic plan with $\mu^+(\chi) = \delta_S$ and χ_n is a sequence of disjoint subtrees with $\mu^+(\chi_n) = m_n \delta_{x_n}$, then $\frac{\mu^-(\chi_n)}{|\mu^-(\chi_n)|} - \delta_{x_n} \to 0$ weakly-$*$ as measures.*

Proof. This is a straightforward consequence of Lemma 8.9.

We can interpret the local irrigation property as a complementary information on the structure of an irrigation network to the one given by Proposition 7.14. This proposition deals with fibers with large flow and says that their structure looks like the structure of a finite irrigation network. The former theorem says that for the rest of the circuit, "small roads don't go far" or in the irrigation interpretation "small tubes irrigate a small neighborhood".

9 The Equivalence of Various Models

This short chapter proves the equivalence of the three mathematical models generalizing the Gilbert-Steiner problem to continuous measures. The proof of this equivalence involves most structure properties proved in Chapter 7. Section 9.1 tells us in short that the Gilbert-Steiner and the optimal traffic plan problems are equivalent for finite atomic measures. Section 9.2 is dedicated to the equivalence of the traffic plan model with the Maddalena et al. *irrigation pattern* model [59]. Finally Section 9.3 treats the equivalence between traffic plans and Xia's *tranport paths*. In all cases the equivalence is strong, as not only the minimal energies are the same but the minimizing objects, of very diverse nature, can be identified with each other. The presentation here follows [13]. Several tools come from [58] and [70].

9.1 Irrigating Finite Atomic Measures (Gilbert-Steiner) and Traffic Plans

Proposition 9.1. *Let π be a transference plan between two finite atomic measures μ^+ and μ^-. An optimal traffic plan for the who goes where problem is a finite graph made of segments. An optimal traffic plan for the irrigation problem between μ^+ and μ^- is a finite graph made of segments with no circuits. Conversely, an optimal graph is an optimal traffic plan.*

Proof. Assume that $\mu^+ = \sum_{i=1}^{k} a_i \delta_{x_i}$ and $\mu^- = \sum_{j=1}^{m} b_j \delta_{y_j}$, $a_i, b_j > 0$. Obviously, there exist traffic plans with finite E^α energy irrigating (μ^+, μ^-) or π. Let χ be an optimum for the who goes where problem. Because of Proposition 7.4, there is an arc $\Gamma_{x_i y_j}$ such that $\chi(\omega, \mathbb{R}^+) = \Gamma_{x_i y_j}$ for all $\omega \in \Omega_{x_i y_j}$ (see Figure 9.1). In addition, two such arcs meet on a single interval, possibly empty. Thus, an optimum for the who goes where problem has the structure of a finite graph. The same argument applies for the irrigation problem. In addition, Proposition 7.8 implies in that case that no circuit occurs. Thus the graph has a tree structure with finitely many branching points. Let us call vertices the sinks, wells and branching points of the graph and edges the pieces of fibers joining two vertices. By optimality the edges are segments.

Conversely consider an optimal irrigation finite graph (G, f). Using Kirchhoff's law, one can by a classical graph flow algorithm decompose

M. Bernot et al., *Optimal Transportation Networks*. Lecture Notes in Mathematics 1955.
© Springer-Verlag Berlin Heidelberg 2009

(G, f) into a finite sum of constant flows on paths, each joining some x_i to some y_j. This decomposition immediately yields a traffic plan structure χ to (G, f). Indeed each path with constant flow is an atomic traffic plan and their sum (or union if they are parameterized) defines a traffic plan χ. By Formula (4.5), $M^\alpha(G) = E^\alpha(\chi)$. Since optimal traffic plans for (μ^+, μ^-) are finite graphs, (G, f) is optimal among traffic plans as well.

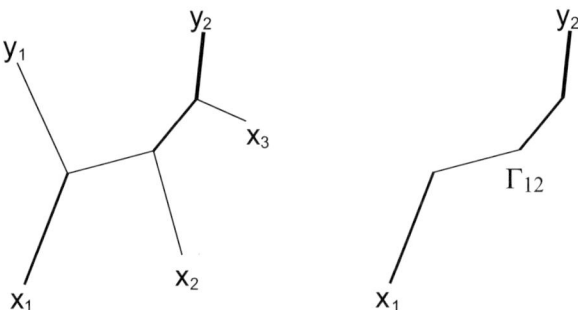

Fig. 9.1. Proposition 7.4 asserts that fibers connecting x_i with y_j follow a single arc Γ_{ij}. Thus, as stated by Proposition 9.1, an optimal traffic plan transporting finite atomic measures has the structure of a finite graph.

9.2 Patterns (Maddalena et al.) and Traffic Plans

This section is dedicated to the equivalence of the traffic plan model with the Maddalena et al. *irrigation pattern* model [59].

Proposition 9.2. *Let $\mu^+ = \delta_S$ be a point source and μ^- be a measure with compact support on $\mathbb{R}^N$. Then the optimal patterns irrigating μ^- from the source point S coincide with the optimal traffic plans.*

Proof. By definition, a pattern is a traffic plan with $\mu^+ = \delta_S$ and conversely a traffic plan with $\mu^+ = \delta_S$ is a pattern. In both cases we can assume that t is the length parameter along the fiber $\chi(\omega, t)$. In [59] the energy is defined by

$$\tilde{E}^\alpha(\chi) = \int_\Omega \int_0^{T(\omega)} |[\omega]_t|^{\alpha-1} d\omega dt$$

where $|[\omega]_t|$ is the measure in Ω of the set of fibers ω' such that $\chi(\omega', s) = \chi(\omega, s)$ for $s \leq t$. In the traffic plan definition,

$$\mathcal{E}^\alpha(\chi) = \int_\Omega \int_0^{T(\omega)} |\chi(\omega, t)|_\chi^{\alpha-1} d\omega dt$$

where $|\chi(\omega, t)|_\chi$ is the measure of the set of fibers passing by $\chi(\omega, t)$. Notice that

$$E^\alpha(\chi) \leq \mathcal{E}^\alpha(\chi) \leq \tilde{E}^\alpha(\chi). \tag{9.1}$$

Let χ be an optimal traffic plan between δ_S and μ^- so that $E^\alpha(\chi) = \mathcal{E}^\alpha(\chi)$. By Proposition 7.11 one can assume without loss of generality that χ is strictly single path. Hence for every ω' for which there is some t' such that $\chi(\omega', t') = \chi(\omega, t)$, one has $\chi(\omega', s) = \chi(\omega, s)$ for $s \leq t = t'$. Thus $|\chi(\omega, t)|_\chi = |[\omega]_t|$ for every (ω, t) and therefore the energies $E^\alpha(\chi)$ and $\tilde{E}^\alpha(\chi)$ are equal. Using (9.1) one concludes that an optimal traffic plan with $\mu^+ = \delta_S$ is an optimal pattern.

Conversely, let χ be an optimal pattern. By (9.1) one can find an optimal traffic plan $\tilde{\chi}$ irrigating μ^-. Since $\tilde{\chi}$ can be assumed to be strictly single path by Proposition 7.11, it satisfies $\tilde{E}^\alpha(\tilde{\chi}) = E^\alpha(\tilde{\chi})$. Thus by (9.1) $E^\alpha(\tilde{\chi}) = \tilde{E}^\alpha(\tilde{\chi}) \geq \tilde{E}^\alpha(\chi) \geq E^\alpha(\chi)$, which proves that χ is an optimal traffic plan.

Remark 9.3. The fact that irrigation patterns are single path has been proven in [79]. Thus the second argument of the above proof is given here for a sake of completeness.

9.3 Transport Paths (Qinglan Xia) and Traffic Plans

Due to the results in Chapter 4, given a traffic plan with finite energy and parameterized by arc-length, $\mathcal{H}^1$ a.e. on $\mathrm{Im}\,\chi$ we may define the tangent vector $\tau_\chi(x)$ which coincides with $\dot{\chi}(\omega, t)$ for almost all fibers ω and times t such that $\chi(\omega, t) = x$. Denote by $C_b(\mathbb{R}^N, \mathbb{R}^N)$ the set of continuous bounded functions from $\mathbb{R}^N$ to $\mathbb{R}^N$. We can consider a traffic plan as a vector measure for every $\varphi \in C_b(\mathbb{R}^N, \mathbb{R}^N)$ by setting, in analogy with (2.1),

$$< \chi, \varphi >:= \int_{\mathbb{R}^N} |x|_\chi \varphi(x) \cdot \tau_\chi(x) \, d\mathcal{H}^1.$$

Then χ_n tends to χ in the sense of vector measures if $< \chi_n, \varphi > \to < \chi, \varphi >$ for every $\varphi \in C_b(\mathbb{R}^N, \mathbb{R}^N)$.

Lemma 9.4. *Let χ be a loop-free traffic plan with finite energy. Then for any $\varphi \in C_b(\mathbb{R}^N, \mathbb{R}^N)$,*

$$< \chi, \varphi >= \int_\Omega \int_0^\infty \varphi(\chi(\omega, t)) \cdot \dot{\chi}(\omega, t) \, d\omega dt.$$

Proof. The proof follows the same steps as the proof of Proposition 4.8 observing that if two fibers $\chi(\omega)$ and $\chi(\omega')$ coincide in a set of times, their tangents coincide $\mathcal{H}^1$-a.e. on the set of coincidence. We shall omit the details.

Proposition 9.5. *Let χ_n, χ be traffic plans such that χ_n are parameterized by arc-length, $\chi_n \to \chi$ with $\mathcal{E}^\alpha(\chi_n) \to \mathcal{E}^\alpha(\chi)$ and $\mathcal{E}^\alpha(\chi) < \infty$. Then $\forall \varphi \in \mathcal{C}(\mathbb{R}^N, \mathbb{R}^N)$,*

$$\int_\Omega \int_0^\infty \varphi(\chi_n(\omega, t)) \cdot \dot{\chi}_n(\omega, t) \, dt d\omega \to \int_\Omega \int_0^\infty \varphi(\chi(\omega, t)) \cdot \dot{\chi}(\omega, t) \, dt d\omega. \quad (9.2)$$

Lemma 9.6. *Let χ_n, χ be traffic plans such that $\chi_n \to \chi$ and $T_{\chi_n}(\omega) \to T_\chi(\omega)$ a.e. where $\int_\Omega T_\chi(\omega) \, d\omega < \infty$. Then (9.2) holds.*

Proof. Let $\omega \in \Omega$ be a fiber such that $T_{\chi_n}(\omega) \to T_\chi(\omega) < \infty$, and $\chi_n(\omega, t) \to \chi(\omega, t)$ uniformly for $t \in [0, L]$ where $L > T(\omega)$. Since $\dot{\chi}_n(\omega, t) \to \dot{\chi}(\omega, t)$ weakly in $L^2([0, L], \mathbb{R}^N)$, we have

$$\int_0^\infty \varphi(\chi_n(\omega, t)) \cdot \dot{\chi}_n(\omega, t) \, dt \to \int_0^\infty \varphi(\chi(\omega, t)) \cdot \dot{\chi}(\omega, t) \, dt, \quad (9.3)$$

for any $\varphi \in \mathcal{C}(\mathbb{R}^N, \mathbb{R}^N)$. Observe that this is true for almost any fiber $\omega \in \Omega$. Since the paths are all contained in a bounded set and $\int_\Omega T_\chi(\omega) d\omega < \infty$, (9.2) follows by the dominated convergence theorem.

Proof. (of Proposition 9.5) By Lemma 9.6, it suffices to prove that $T_{\chi_n}(\omega) \to T_\chi(\omega)$ almost everywhere. We already know by Lemma 3.20 that $T_\chi(\omega) \leq \liminf_n T_{\chi_n}(\omega)$ almost everywhere. Since $(\chi_n(\omega))_n$ are parameterized by arc-length until they stop, by discarding a null set of $\Omega \times \mathbb{R}^+$ we may assume that (ω, t) is not in $\bigcup_n \{(\omega, T_{\chi_n}(\omega)) : \omega \in \Omega\} \cup \{(\omega, T_\chi(\omega)) : \omega \in \Omega\}$ and $\liminf_n |\dot{\chi}_n(\omega, t)|$ is either 0 or 1. In case that is 1, we have

$$|\dot{\chi}(\omega, t)| \leq \liminf_n |\dot{\chi}_n(\omega, t)|. \quad (9.4)$$

In case that is 0, there is a subsequence n_j such that $\dot{\chi}_{n_j}(\omega, t) = 0$, hence $t \geq T_{\chi_{n_j}}(\omega)$. Then $T_\chi(\omega) \leq \liminf T_{\chi_{n_j}}(\omega) \leq t$ and (9.4) still holds. By Proposition 3.24 we also have

$$\limsup_n |\chi_n(\omega, t)|_{\chi_n} \leq |\chi(\omega, t)|_\chi. \quad (9.5)$$

Thus

$$|\chi(\omega, t)|_\chi^{\alpha - 1} |\dot{\chi}(\omega, t)| \leq \liminf_n |\chi_n(\omega, t)|_{\chi_n}^{\alpha - 1} |\dot{\chi}_n(\omega, t)| \qquad \text{a.e. on } \Omega \times [0, \infty).$$

Thus, if for simplicity we denote $T(\omega) = T_\chi(\omega)$, $T_n(\omega) = T_{\chi_n}(\omega)$,

$$g_n(\omega, t) = |\chi_n(\omega, t)|_{\chi_n}^{\alpha - 1} |\dot{\chi}_n(\omega, t)|, \qquad g(\omega, t) = |\chi(\omega, t)|_\chi^{\alpha - 1} |\dot{\chi}(\omega, t)|,$$

we have just proven that

$$g(\omega, t) \leq \liminf_n g_n(\omega, t) \quad (9.6)$$

Since $g(\omega, t) = 0$ when $t > T(\omega)$, one has

$$\mathcal{E}^\alpha(\chi_n) = \int_\Omega \int_0^{T_n(\omega)} g_n(\omega, t) dt d\omega$$

$$= \int_\Omega \int_0^{T(\omega)} g_n \, dt d\omega + \int_\Omega \int_{T(\omega)}^{\max(T(\omega), T_n(\omega))} g_n \, dt d\omega,$$

$$\mathcal{E}^\alpha(\chi) = \int_\Omega \int_0^{T(\omega)} g(\omega, t) dt d\omega.$$

Thus,

$$\mathcal{E}^\alpha(\chi) = \lim_n \mathcal{E}^\alpha(\chi_n)$$

$$\geq \liminf_n \int_\Omega \int_0^{T(\omega)} g_n \, dt d\omega + \limsup_n \int_\Omega \int_{T(\omega)}^{\max(T(\omega), T_n(\omega))} g_n \, dt d\omega,$$

and by Fatou's lemma and (9.6),

$$\mathcal{E}^\alpha(\chi) \geq \mathcal{E}^\alpha(\chi) + \limsup_n \int_\Omega \int_{T(\omega)}^{\max(T(\omega), T_n(\omega))} g_n \, dt d\omega.$$

Now, for $t \in [0, T_n(\omega)]$, $g_n(\omega, t) = |\chi_n(\omega, t)|_{\chi_n}^{\alpha-1} |\dot\chi_n(\omega, t)| \geq 1$. Thus

$$0 \geq \limsup_n \int_\Omega \int_{T(\omega)}^{\max(T(\omega), T_n(\omega))} dt d\omega = \limsup_n \int_\Omega (T_n(\omega) - T(\omega))^+ d\omega.$$

Thus, $T(\omega) \geq \limsup_n T_n(\omega)$ a.e. and therefore $T_{\chi_n}(\omega) \to T_\chi(\omega)$ almost everywhere.

We recall that the definition of $M^\alpha(\mu(\chi))$ has been given in Section 2.3.1 of Chapter 2. Thus, as a consequence of Proposition 7.14, Lemma 9.4 and Proposition 9.5 we have:

Corollary 9.7. *Let χ be an optimal traffic plan of finite E^α energy with $\mu(\chi) = (\mu^+, \mu^-)$. Then*

$$E^\alpha(\chi) = E^\alpha(\mu(\chi)) \geq M^\alpha(\mu(\chi)). \tag{9.7}$$

Proof. Notice that $\mathcal{E}^\alpha(\chi) = E^\alpha(\chi)$. We may assume that χ is parameterized by arc-length and recall that χ is equivalent to a loop-free traffic plan. Hence, we may assume that it has simple paths. Let χ_n be the sequence of traffic plans determined in Proposition 7.14 which are restrictions of χ and such that $\chi_n \to \chi$ and $E^\alpha(\chi_n) = \mathcal{E}^\alpha(\chi_n) \to E^\alpha(\chi)$. Then by Lemma 9.4 and Proposition 9.5 we know that χ_n converges to χ in the sense of vector measures. The inequality (9.7) follows from this.

Proposition 9.8. *Let (μ^+, μ^-) be two Radon measures of equal mass. Then*

$$E^\alpha(\mu^+, \mu^-) = M^\alpha(\mu^+, \mu^-). \tag{9.8}$$

Moreover, if these energies are finite, the optima are the same.

Proof. Assume that $M^\alpha(\mu^+, \mu^-) < \infty$. By Theorem 3.1 in [94] there is a transport path T from μ^+ to μ^- such that $M^\alpha(T) = M^\alpha(\mu^+, \mu^-)$. Let μ_n^+, μ_n^- be finite atomic measures of equal mass, G_n finite graphs (which we may assume to be loop-free) such that $G_n \rightharpoonup T$ in the sense of vector measures, $\mu_n^+ \rightharpoonup \mu^+$, $\mu_n^- \rightharpoonup \mu^-$, and $\mathcal{E}^\alpha(G_n) = E^\alpha(G_n) = M^\alpha(G_n) \to M^\alpha(T)$. By the compactness Theorem 3.28 for traffic plans, there is a traffic plan χ and a subsequence $G_n \to \chi$ such that

$$E^\alpha(\chi) \leq \liminf_n E^\alpha(G_n) = M^\alpha(\mu^+, \mu^-)$$

and χ irrigates (μ^+, μ^-). Since $E^\alpha(\mu^+, \mu^-) \leq E^\alpha(\chi)$, we obtain

$$E^\alpha(\mu^+, \mu^-) \leq M^\alpha(\mu^+, \mu^-). \tag{9.9}$$

Since the inequality (9.9) holds also when $M^\alpha(\mu^+, \mu^-) = \infty$, it holds in any case. If $E^\alpha(\mu^+, \mu^-) = \infty$, then (9.8) holds. Thus, we may assume that $E^\alpha(\mu^+, \mu^-) < \infty$. In that case, by Corollary 9.7, we have $M^\alpha(\mu^+, \mu^-) \leq E^\alpha(\mu^+, \mu^-)$ and (9.8) holds.

Assume that the energies in (9.8) are finite. Now, since we are assuming that G_n are loop-free we may choose χ_n to be a loop-free traffic plan parameterizing G_n, so that $\mathcal{E}^\alpha(\chi_n) = E^\alpha(\chi_n) = M^\alpha(G_n)$. We may assume that χ_n are parameterized by arc-length. Moreover, we know that the traffic plan χ, limit of χ_n, is loop-free, up to a set of fibers with zero measure. Hence, by Lemma 9.4 and Proposition 9.5 we know that χ_n converges to χ in the sense of vector measures. Thus, χ can be considered as a parameterization of T. The fact that any optimal traffic plan can be considered as an optimum transport path is contained in the proof of Corollary 9.7.

The equivalence of the irrigation and the Qinglan Xia's model has also been proved in [58]. In this paper a very general axiomatic formulation of the transport costs and their derived distances is proposed and the above equivalence results are particular cases of the general setting.

9.4 Optimal Transportation Networks as Flat Chains

The optimization of transportation networks has also been studied in [70] using the formalism of currents [40]. In this context the flow is modeled as a one dimensional current, more precisely, by a one-dimensional flat chain (one-dimensional currents that are approximable in the flat norm by polyhedral

chains, which are those that can be written as finite linear combinations
of segments). Moreover, the authors proved the equivalence of the current
formulation with the traffic plan formulation, a concept known as transport
in [70]. Indeed, to any traffic plan P the authors associate a normal one-
dimensional current (i.e. a current with finite mass whose boundary has also
finite mass) whose boundary coincides with $\mu^+(P) - \mu^-(P)$. Conversely, any
normal one-dimensional current with compact support can be represented as
a traffic plan. We include a simplified version of their results. For that, we
recall first some basic definitions on currents.

If U is an open set in $\mathbb{R}^N$ and $k \in \mathbb{N}$, let $\mathcal{E}^k(U)$, resp. $\mathcal{D}^k(U)$, be the
vector space of C^∞-differential forms of degree k, resp. C^∞-differential forms
of degree k with compact support in U. Both vector spaces are endowed with
the topology of uniform convergence of the forms and its derivatives of any
order on compact sets of U. A k-dimensional current $\mathbb{T}$ on U is a continuous
linear form on $\mathcal{D}^k(U)$. A k-dimensional current with compact support in U
are those currents that can be extended as linear forms on $\mathcal{E}^k(U)$. If $\mathbb{T}$ is a
k-dimensional current in U and $V \subseteq U$ is an open set, we define

$$\mu_{\mathbb{T}}(V) := \sup\{\mathbb{T}(\omega) : \operatorname{supp}\omega \subset V, \quad \|\omega\|_\infty \leq 1\}.$$

We define the mass of $\mathbb{T}$ as $\mathbb{M}(\mathbb{T}) := \mu_{\mathbb{T}}(U)$. When $\mathbb{M}(\mathbb{T}) < \infty$, then $\mu_{\mathbb{T}}$
defines a finite Radon measure and $\mathbb{T}$ is representable as

$$\mathbb{T}(\omega) = \int_U \langle \tau_{\mathbb{T}}(x), \omega(x) \rangle \, d\mu_{\mathbb{T}}(x), \qquad \omega \in \mathcal{D}^m(U),$$

where $\tau_{\mathbb{T}}$ is a unit simple k-vector field [40]. If $k \geq 1$ and $\mathbb{T}$ is a k-dimensional
current in U, we define the boundary of $\mathbb{T}$ as the $k-1$-dimensional current
$\partial\mathbb{T}$ given by

$$\partial\mathbb{T}(\omega) = \mathbb{T}(d\omega) \qquad \omega \in \mathcal{D}^{k-1}(U),$$

where $d\omega$ denotes the exterior derivative of ω, which is a k-form. We say that
a k-dimensional current $\mathbb{T}$ in U is normal if $\mathbb{M}(\mathbb{T}) < \infty$ and, in case that
$k \geq 1$, $\mathbb{M}(\partial\mathbb{T}) < \infty$.

The flat norm of the current $\mathbb{T}$ with support in a compact subset $Q \subset U$
is given by

$$\mathbb{F}_Q(\mathbb{T}) := \inf\{\mathbb{M}(\mathbb{T}-\partial\mathbb{S})+\mathbb{M}(\mathbb{S}) : \mathbb{S} \text{ is a } k+1\text{-dim. current with supp. in } Q\}.$$

We say that a k-dimensional current $\mathbb{T}$ with support in the compact set Q
is a flat chain if there is a sequence of normal currents with support in Q
converging to $\mathbb{T}$ in the flat norm $\mathbb{F}_Q$. We recall that if $X \subset \operatorname{int}(C) \subset U$, X and
C being compact, and $\mathbb{T}$ is a k-dimensional current with support in X, then $\mathbb{T}$
can be approximated in the flat norm $\mathbb{F}_C$ by k-dimensional polyhedral chains
with support in C [40], Theorem 4.1.23. A k-dimensional polyhedral chain is
a finite linear combination (with real coefficients) of oriented simplexes.

To any 1-Lipschitz curve $\gamma \in K$ we associate the one-dimensional current $[\gamma]$ defined by

$$[\gamma](\omega) = \int_{\mathbb{R}+} \langle \dot{\gamma}(t), \omega(\gamma(t)) \rangle \, dt.$$

Observe that

$$\mathbb{M}([\gamma]) \leq \int_{\mathbb{R}+} |\dot{\gamma}(t)| \, dt.$$

The next two results were proved in [70].

Proposition 9.9. *Let P be a traffic plan in K such that $\int_K T(\gamma) \, dP < \infty$. Then*

$$\mathbb{T}_P(\omega) := \int_K [\gamma](\omega) \, dP(\gamma)$$

defines a normal one-dimensional current on $\mathbb{R}^N$ with compact support such that

$$\partial \mathbb{T}_P = \mu^+(P) - \mu^-(P).$$

In particular, if $\mu^+(P) \wedge \mu^-(P) = 0$, then $(\partial \mathbb{T}_P)^+ = \mu^+(P)$, $(\partial \mathbb{T}_P)^- = \mu^-(P)$.

Proof. We have to prove that $\mathbb{T}_P$ is a continuous linear form on $\mathcal{E}^1(\mathbb{R}^N)$, and both $\mathbb{T}_P$ and its boundary have finite mass. From its definition it is clear that $\mathbb{T}_P$ is a current with compact support in $\mathbb{R}^N$. Using the definition of mass, we clearly have

$$\mathbb{M}(\mathbb{T}_P) \leq \int_K \mathbb{M}([\gamma]) \, dP(\gamma) \leq \int_K T(\gamma) \, dP < \infty.$$

To compute the boundary, let $f \in C^\infty(\mathbb{R}^N)$. Then

$$\partial \mathbb{T}_P(f) = \mathbb{T}_P(df) = \int_K \int_0^{T(\gamma)} \langle \nabla f(\gamma(t)), \dot{\gamma}(t) \rangle \, dt \, d\mathbb{T}_P(\gamma)$$

$$= \int_K \int_0^{T(\gamma)} \frac{d}{dt}(f \circ \gamma)(t) \, dt \, d\mathbb{T}_P(\gamma)$$

$$= \int_K (f(\gamma(T(\gamma))) - f(\gamma(0))) \, d\mathbb{T}_P(\gamma)$$

$$= \int_{\mathbb{R}^N} f(x) \, d\mu^+(P) - \int_{\mathbb{R}^N} f(x) \, d\mu^-(P).$$

This implies that $\partial \mathbb{T}_P = \mu^+(P) - \mu^-(P)$ which has finite mass.

Proposition 9.10. *Given a one-dimensional real flat chain $\mathbb{T}$ with compact support and finite mass $\mathbb{M}(\mathbb{T}) < \infty$, there exists a traffic plan P such that $\mathbb{T} = \mathbb{T}_P$ and*

$$\mathbb{M}(\mathbb{T}) = \int_K \mathbb{M}([\gamma]) \, dP(\gamma) = \int_K L([\gamma]) \, dP(\gamma). \tag{9.10}$$

Observe that this is a weaker statement than required since we are not saying that $(\partial\mathbb{T})^+ = \mu^+(\boldsymbol{P})$ and $(\partial\mathbb{T})^- = \mu^-(\boldsymbol{P})$ but only that $\partial\mathbb{T} = \mu^+(\boldsymbol{P}) - \mu^-(\boldsymbol{P})$. In case that $\mathbb{T}$ is acyclic, i.e. if $\partial\mathbb{T} = 0$, then we can choose $\boldsymbol{P}$ so that this holds, but this requires a more technical proof and we refer to [70] for it.

Proof. Step 1. The assertions of the theorem hold when $\mathbb{T}$ is a one-dimensional real polyhedral chain. Indeed, in this case $\mathbb{T}$ can be written as a finite sum $\mathbb{T} = \sum_m \theta_m \mathbb{T}_m$ where $\theta_m > 0$ are real multiplicities and $\mathbb{T}_m$ are currents associated to the segments $[a_m, b_m]$, $a_m, b_m \in \mathbb{R}^N$. Let $\gamma_m(t) = (1-t)a_m + tb_m$, $t \in [0,1]$ be a parameterization of $[a_m, b_m]$ and let

$$\boldsymbol{P} := \sum_m \theta_m \delta_{\gamma_m},$$

where δ_{γ_m} is the Dirac measure concentrated on γ_m. Then we have

$$\mathbb{T}(\omega) = \sum_m \theta_m \int_0^{T(\gamma_m)} \langle \omega(\gamma_m(t)), b_m - a_m \rangle \, dt$$

$$= \sum_m \theta_m [\gamma_m](\omega) = \int_K [\gamma](\omega) \, d\boldsymbol{P},$$

that is, $\mathbb{T} = \mathbb{T}_{\boldsymbol{P}}$. By the definition of $\boldsymbol{P}$, we have $\mathbb{M}([\gamma]) = L(\gamma)$ for $\boldsymbol{P}$-a.e. $\gamma \in K$, hence

$$\mathbb{M}(\mathbb{T}) = \sum_m \theta_m |b_m - a_m| = \int_K \mathbb{M}([\gamma]) \, d\boldsymbol{P}(\gamma) = \int_K L(\gamma) \, d\boldsymbol{P}(\gamma).$$

Moreover, since the support of $\boldsymbol{P}$ is formed by the curves γ_m, we may take X as a compact set containing the support of $\mathbb{T}$ and take K as the 1-Lipschitz curves on X. Then $\boldsymbol{P}$-a.e. $\gamma \in K$ is a segment contained in the support of $\mathbb{T}$ and $\boldsymbol{P}$ is concentrated in $\{\gamma \in K : \mathcal{H}^1(\gamma) \leq \operatorname{diam} \operatorname{supp} \mathbb{T}\}$.

Observe at this point that we are not saying that $(\partial\mathbb{T})^+ = \mu^+(\boldsymbol{P})$ and $(\partial\mathbb{T})^- = \mu^-(\boldsymbol{P})$ but only that $\partial\mathbb{T} = \mu^+(\boldsymbol{P}) - \mu^-(\boldsymbol{P})$.

Step 2. Assume that $\mathbb{T}$ is any one-dimensional real flat chain with compact support and finite mass. Let $\mathbb{T}_n$ be a sequence of one dimensional polyhedral chains which converge to $\mathbb{T}$ in the flat norm and $\mathbb{M}(\mathbb{T}_n) \to \mathbb{M}(\mathbb{T})$ [40]. Moreover, we may assume that the supports of $\mathbb{T}_n$ are all contained in a compact set X of $\mathbb{R}^N$ [40]. By Step 1 we find a traffic plan $\boldsymbol{P}_n$ in K such that

$$\mathbb{T}_n = \mathbb{T}_{\boldsymbol{P}_n} \qquad \text{and} \qquad \mathbb{M}(\mathbb{T}_n) = \int_K L(\gamma) \, d\boldsymbol{P}_n(\gamma).$$

Now, by extracting a subsequence, we may assume that $\boldsymbol{P}_n$ converges to a traffic plan $\boldsymbol{P}$. Moreover, by Step 1 we know that the support of $\boldsymbol{P}_n$ is concentrated in $\{\gamma \in K : \mathcal{H}^1(\gamma) \leq \operatorname{diam}(X)\}$. As in the proof of Lemma 9.6, we can check that for any smooth 1-form ω in $\mathbb{R}^N$, the function $\gamma \in K \to [\gamma](\omega)$ is continuous. Then

$$\mathbb{T}_{\boldsymbol{P}_n}(\omega) = \int_K [\gamma](\omega) \, d\boldsymbol{P}_n \to \int_K [\gamma](\omega) \, d\boldsymbol{P}$$

as $n \to \infty$ for any smooth 1-form ω. Hence $\mathbb{T} = \mathbb{T}_{\boldsymbol{P}}$. Now, we observe that the function $\gamma \in K \mapsto \mathbb{M}([\gamma])$ is a lower semicontinous function. Hence

$$\int_K \mathbb{M}([\gamma]) \, d\boldsymbol{P}(\gamma) \le \liminf_n \int_K \mathbb{M}([\gamma]) \, d\boldsymbol{P}_n(\gamma) = \lim_n \mathbb{M}(\mathbb{T}_n) = \mathbb{M}(\mathbb{T}).$$

Since we always have that $\mathbb{M}(\mathbb{T}) \le \int_K \mathbb{M}([\gamma]) \, d\boldsymbol{P}(\gamma)$ we obtain the first equality in (9.10). Similarly

$$\int_K L([\gamma]) \, d\boldsymbol{P}(\gamma) \le \liminf_n \int_K L([\gamma]) \, d\boldsymbol{P}_n(\gamma) = \lim_n \mathbb{M}(\mathbb{T}_n) = \mathbb{M}(\mathbb{T}).$$

Since we have that

$$\mathbb{M}(\mathbb{T}) = \int_K \mathbb{M}([\gamma]) \, d\boldsymbol{P}(\gamma) \le \int_K L([\gamma]) \, d\boldsymbol{P}(\gamma),$$

we have the second equality in (9.10).

We call $\mathbb{T}$ a rectifiable current if there is an $\mathcal{H}^k$ countably rectifiable set $\Sigma \subset \mathbb{R}^N$ and a function $\theta \in L^1(\mathcal{H}^k|_\Sigma)$, called the multiplicity of $\mathbb{T}$, such that

$$\mathbb{T}(\omega) = \int_\Sigma \theta(x) \langle \tau_{\mathbb{T}}(x), \omega(x) \rangle \, d\mathcal{H}^k(x),$$

where $\tau_{\mathbb{T}} : \Sigma \to \mathbb{R}^N$ is a unit simple k-vector such that $\tau_{\mathbb{T}}(x)$ defines the approximate tangent plane to Σ at x, $\mathcal{H}^k$-a.e. $x \in \Sigma$. We call $\tau_{\mathbb{T}}$ an orientation of Σ. In this case, we shall write $\mathbb{T} = \theta[\Sigma]$. One can show that if $\mathbb{T}$ is a flat chain with finite mass $\mathbb{M}(\mathbb{T}) < \infty$, then $\mathbb{T}$ is a rectifiable current if and only if for some $\mathcal{H}^k$ countably rectifiable set $\Sigma \subset \mathbb{R}^N$ we have that $\mu_{\mathbb{T}} = \mu_{\mathbb{T}}|_\Sigma$ [40]. Given a k-dimensional rectifiable current $\mathbb{T} = \theta[\Sigma]$ and $\alpha \in [0,1]$ we define the α-mass of $\mathbb{T}$ by the formula

$$\mathbb{M}^\alpha(\mathbb{T}) = \int_\Sigma \theta(x)^\alpha \, d\mathcal{H}^k(x).$$

For one-dimensional rectifiable currents, this is the analog of the energy E^α. The functional M^α is lower semicontinuous on rectifiable currents with respect to the flat norm convergence. Hence, it can be extended to a lower semicontinuous functional defined on flat chains.

The following result was proved by White in [91].

Theorem 9.11. *Let $\mathbb{T}$ be a current such that $\mathbb{M}(\mathbb{T}) < \infty$ and $\mathbb{M}^\alpha(\mathbb{T}) < \infty$ for some $\alpha \in [0,1)$. Then $\mathbb{T}$ is rectifiable.*

10 Irrigability and Dimension

Let μ be a positive Borel measure on $\mathbb{R}^N$ which we assume without loss of generality to have total mass 1. If $E^\alpha(\delta_S, \mu) < \infty$ for some $\alpha \in [0, 1]$, then we say that μ is irrigable with respect to α. In that case, notice that μ is also β-irrigable for $\beta > \alpha$. This observation proves the existence of a critical exponent α associated with μ and defined as the smallest exponent such that μ is α-irrigable. The aim of the chapter is to link this exponent to more classical dimensions associated with μ such as the Hausdorff and Minkowski dimensions of the support of μ. Inequalities between these dimensions and an "irrigability dimension" will be established. A striking result is that when μ is Ahlfors regular, all considered dimensions are equal. To illustrate the results, let us take the case where μ is the Lebesgue measure of a ball in dimension N. We already know in that case that μ is irrigable for every $\alpha > 1 - \frac{1}{N}$. What happens if α is critical? Corollary 10.16 gives the answer: if any probability measure μ with a bounded support is α-irrigable, then $\alpha > \frac{1}{N'}$. Thus the N-dimensional Lebesgue measure is not $1 - \frac{1}{N}$ irrigable. Here the presentation and results follow closely Devillanova's PhD [28] and Devillanova-Solimini [78].

We shall consider only traffic plans with a point source S, that is, patterns in the sense of Maddalena et al. [59]. Since a measure irrigable from S is irrigable from any other point, we shall omit the mention of S. For a sake of simpler notation we shall note $\mu_\chi = \mu^+(\chi)$.

10.1 Several Concepts of Dimension of a Measure and Irrigability Results

Definition 10.1. *For a fixed real number $\alpha \in]0, 1[$ we shall call* critical dimension *of the exponent α the constant $d_\alpha = \frac{1}{1-\alpha} = \left(\frac{1}{\alpha}\right)' > 1$.*

Definition 10.2. *Let $\alpha \in]0, 1[$ be given and let μ be a probability measure on $\mathbb{R}^N$. We shall say that μ is an irrigable measure with respect to α (or that μ is α-irrigable) if there exists a pattern with finite cost $E^\alpha(\chi) < +\infty$ such that $\mu_\chi = \mu$.*

It is clear that two approaches are possible and equivalent: one can fix a constant $\alpha \in]0, 1[$ and investigate the irrigable measures with respect to this constant or fix a measure μ and find out the constants $\alpha \in]0, 1[$ with respect

M. Bernot et al., *Optimal Transportation Networks*. Lecture Notes in Mathematics 1955.
© Springer-Verlag Berlin Heidelberg 2009

to which μ is irrigable. This second point of view leads us to introduce the following definition.

Definition 10.3. *Let μ be a positive Borel measure on $\mathbb{R}^N$, then we shall call* irrigability dimension *of μ the number*

$$d(\mu) = \inf\{d_\alpha : \mu \text{ is irrigable with respect to } \alpha\} \ .$$

Remark 10.4. For any probability measure μ, by definition, the irrigability dimension $d(\mu)$ of μ is larger or equal to 1.

Remark 10.5. If μ is an irrigable measure with respect to α, then μ is also irrigable with respect to every constant $\beta \in [\alpha, 1[$. Indeed, $E^\beta(\chi) \leq E^\alpha(\chi)$.

Remark 10.6. By the definition of $d(\mu)$ and by Remark 10.5 it follows that for a given $\alpha \in]0, 1[$ and for a given measure μ:

1. if $d(\mu) < d_\alpha$ then μ is α-irrigable;
2. if $d(\mu) > d_\alpha$ then μ is not α-irrigable.

As we shall show in Section 10.4, both cases can occur when $d_\alpha = d(\mu)$, see Examples 10.33 and 10.35. Our first aim is to give estimates of $d(\mu)$ in terms of geometrical properties of the measure μ.

Definition 10.7. *We shall say that a positive Borel measure μ on $\mathbb{R}^N$ is* concentrated on *a Borel set B if $\mu(\mathbb{R}^N \backslash B) = 0$ and we shall call* concentration dimension *of μ the smallest Hausdorff dimension $d(B)$ of a set B on which μ is concentrated i.e. the number*

$$d_c(\mu) = \inf\{d(B) : \mu \text{ is concentrated on } B\} \ .$$

Definition 10.8. *We shall denote by $\operatorname{supp}(\mu)$ the support of μ in the sense of distributions and shall call* support dimension *of μ, $d_s(\mu)$, its Hausdorff dimension.*

Remark 10.9. The support of a measure can be characterized as the smallest closed set on which μ is concentrated and the existence of such a set *a priori* follows from the separability of $\mathbb{R}^N$, precisely from the Lindelöf property. While, as stated above, the existence of the smallest closed set on which μ is concentrated is granted, it is clear that the smallest set on which μ is concentrated, in general, does not exist. This is the reason for which an infimum is taken in Definition 10.7. Since $\operatorname{supp}(\mu)$ is a set on which μ is concentrated, it follows that

$$d_c(\mu) \leq d_s(\mu) \ .$$

These two geometrical dimensions are not sufficient to study the irrigability of a measure, as we shall show later in examples 10.28 and 10.32.

Definition 10.10. *Let $X \subset \mathbb{R}^N$ be a bounded set. We shall call Minkowski dimension of the set X (see [60]) the constant*

$$d_M(X) = N - \liminf_{\delta \to 0} \log_\delta |N_\delta(X)| \tag{10.1}$$

where, for all $\delta > 0$,

$$N_\delta(X) = \{ y \in \mathbb{R}^N | d(y, X) < \delta \} .$$

Clearly,

$$0 \le d_M(X) \le N \qquad \forall X \ne \emptyset . \tag{10.2}$$

The Minkowski dimension of a set $X \subset \mathbb{R}^N$ can be characterized by the following two properties:

$$\forall \beta < d_M(X) \quad \limsup_{\delta \to 0} |N_\delta(X)| \delta^{\beta - N} = +\infty \tag{10.3}$$

and

$$\forall \beta > d_M(X) \quad \lim_{\delta \to 0} |N_\delta(X)| \delta^{\beta - N} = 0 . \tag{10.4}$$

Lemma 10.11. *Let $X \subset \mathbb{R}^N$.*

a) If $\beta > d_M(X)$, then we can cover X by using $\delta^{-\beta}$ balls of radius δ for all δ sufficiently small.

b) If $\beta < d_M(X)$, then it is not possible to find a constant $C > 0$ such that one can cover X with only $C\delta^{-\beta}$ balls of radius δ for all δ sufficiently small. As a consequence, $d_M(X)$ coincides with

$$\inf\{\beta \ge 0 : X \text{ can be covered by } C_\beta \delta^{-\beta} \text{ balls of radius } \delta \text{ for all } \delta \le 1\}. \tag{10.5}$$

Proof. a) If $\beta > d_M(X)$, by (10.4), for all $C > 0$ and for $\delta > 0$ sufficiently small

$$|N_{\frac{\delta}{2}}(X)| \le C\delta^{N - \beta} .$$

We consider any family of disjoint balls $(B_i)_{i \in I}$ of radius $\frac{\delta}{2}$ contained in $N_{\frac{\delta}{2}}(X)$. Since $|B_i| = b_N \left(\frac{\delta}{2} \right)^N$, for all $i \in I$ (b_N stands for the measure of the unit ball of $\mathbb{R}^N$), we know that $\text{card}(I) b_N \left(\frac{\delta}{2} \right)^N \le |N_{\frac{\delta}{2}}(X)| \le C\delta^{N - \beta}$. Taking for C the constant $\frac{b_N}{2^N}$ we obtain

$$\text{card}(I) \le C \frac{2^N}{b_N} \delta^{-\beta} = \delta^{-\beta} . \tag{10.6}$$

Thus the cardinality of any family of disjoint balls contained in $N_{\frac{\delta}{2}}(X)$ is bounded by $\delta^{-\beta}$. Let us take a maximal family of such balls. Then the family of balls with the same centers but with double radius is a covering of X. Inequality (10.6) gives the thesis.

b) Assume by contradiction that there exists a constant $C > 0$ such that for every δ small enough X can be covered by $C\delta^{-\beta}$ balls with radius δ. Doubling the radius of these balls we get a covering of $N_\delta(X)$, so we have

$$|N_\delta(X)| \le C\delta^{-\beta} b_N(2\delta)^N \le const\ \delta^{N-\beta}\ ,$$

which gives $\beta \ge d_M(X)$ by (10.3).

Definition 10.12. *Let μ be a probability measure. We shall call* Minkowski dimension *of μ the number*

$$d_M(\mu) = \inf\{d_M(X) : \mu \text{ is concentrated on } X\}. \tag{10.7}$$

Remark 10.13. For any subset X of $\mathbb{R}^N$ the Minkowski dimensions of X and of its closure $\overline{X}$ are the same. Therefore for any probability measure μ we have

$$d_M(\mu) = d_M(\text{supp}(\mu))\ .$$

Moreover the Hausdorff dimension $d(X)$ of a set X is less or equal than $d_M(X)$. So for any probability measure μ

$$d_s(\mu) \le d_M(\mu)\ . \tag{10.8}$$

Collecting Remark 10.9 and (10.8) we obtain

$$d_c(\mu) \le d_s(\mu) \le d_M(\mu)\ . \tag{10.9}$$

As indicated by the next theorem a similar estimate is true for $d(\mu)$.

Theorem 10.14. *(Lower and Upper bound on $d(\mu)$)*

$$d_c(\mu) \le d(\mu) \le \max\{d_M(\mu), 1\}\ . \tag{10.10}$$

The first inequality in (10.10) is a straightforward consequence of a deeper and more precise result stated in the following theorem, whose proof will be given in Section 10.2.

Theorem 10.15. *Let $\alpha \in]0, 1[$ and let μ be an α-irrigable probability measure, then μ is concentrated on a d_α-negligible set, in particular,*

$$d_c(\mu) \le d_\alpha\ . \tag{10.11}$$

Theorems 10.14 and 10.15 widely answer the question considered in [94] about the values of α which make every measure of bounded support irrigable. Indeed, we can deduce the following corollaries.

Corollary 10.16. *Let* $\alpha \in]0, 1[$. *If* $\alpha > \frac{1}{N'}$, *then any probability measure* μ *with a bounded support is* α-*irrigable. Conversely, if any probability measure* μ *with a bounded support is* α-*irrigable, then* $\alpha > \frac{1}{N'}$.

Proof. Remarking that $\alpha > \frac{1}{N'}$ is equivalent to $d_\alpha = \left(\frac{1}{\alpha}\right)' > N$, combining (10.2) with (10.10), we have for every μ

$$d_\alpha > N \geq \max\{d_M(\mu), 1\} \geq d(\mu) ,$$

so every probability measure μ with a bounded support is α-irrigable by Remark 10.6,1.

To prove the converse assertion, observe that from Theorem 10.15 we have that any probability measure μ with a bounded support is concentrated on a d_α-negligible set. So, $N < d_\alpha$, namely $\alpha > \frac{1}{N'}$.

In spite of inequalities (10.10) and (10.9) it is not possible to establish some general inequality between $d(\mu)$ and $d_s(\mu)$, as shown in Section 10.4 by the examples 10.28 and 10.32. By the following lemmas we shall make the estimates on the dimension $d(\mu)$ more precise in the case where the probability measure μ has some uniformity property.

Definition 10.17. *Let* μ *be a probability measure and* $\beta \geq 0$. *We shall say that* μ *is Ahlfors regular in dimension* β *if*

$$(AR) \qquad \exists C_1, C_2 > 0 \; s.t. \; \forall r \in [0, 1], \; \forall x \in \text{supp}(\mu) :$$

$$\tag{10.12}$$

$$C_1 r^\beta \leq \mu(B(x, r)) \leq C_2 r^\beta .$$

We shall separately consider the two bounds in (AR). So for a probability measure μ and a real number $\beta \geq 0$ we shall consider the two conditions

$$(LAR) \; \exists C > 0 \; \text{s.t.} \; \forall r \in [0, 1], \; \forall x \in \text{supp}(\mu) : \; C r^\beta \leq \mu(B(x, r)) , \tag{10.13}$$

and

$$(UAR) \; \exists C > 0 \; \text{s.t.} \; \forall r \in [0, 1], \; \forall x \in \text{supp}(\mu) : \; \mu(B(x, r)) \leq C r^\beta . \tag{10.14}$$

In (UAR) the restriction $x \in \text{supp}(\mu)$ can be removed by multiplying C_2 by 2^β.

Definition 10.18. *A probability measure* μ *satisfies the uniform density property (in short u.d.p.) in dimension* $\beta \geq 0$ *on a set* M *if*

$$\exists C_1 > 0 \; s.t. \; \forall x \in M, \; \forall r \in [0, 1] : \; C_1 r^\beta \leq \mu(B(x, r)) .$$

Lemma 10.19. *Let* ν *be a probability measure which satisfies the u.d.p. in dimension* $\beta \geq 0$ *on a subset* B. *Then*

$$d_M(B) \leq \beta . \tag{10.15}$$

In particular, if μ is a probability measure and $\beta \geq 0$ is such that μ satisfies (LAR) (i.e. μ satisfies the uniform density property in dimension β on supp(μ)*), then*

$$d_M(\mu) \leq \beta . \tag{10.16}$$

Proof. Let us fix $\delta > 0$ and let us consider any family $(B_i)_{i \in I}$ of disjoint balls of radius $\frac{\delta}{2}$ with centers on B. By assumption, $\nu(B_i) \geq C2^{-\beta}\delta^\beta$ and $\nu(B) \leq 1$, therefore card$(I) \leq 2^\beta C^{-1}\delta^{-\beta}$. So we can consider a family $(B_i)_{i \in I}$ as above maximal by inclusion. The maximality of $(B_i)_{i \in I}$ guarantees that, for any other point $x \in B$, $d(x, \bigcup_{i \in I} B_i) < \frac{\delta}{2}$ holds. Therefore the family $(\tilde{B}_i)_{i \in I}$ obtained by doubling the radii of the balls B_i is a covering of B. Thus B can be covered by $C\delta^{-\beta}$ balls with radius δ arbitrarily small. By Lemma 10.11 we conclude that $d_M(B) \leq \beta$.

Remark 10.20. The inequality (10.16) still holds if one assumes the existence of a probability measure ν which satisfies the uniform density property in dimension β on a set B on which μ is concentrated.

Lemma 10.21. *Let μ be a probability measure concentrated on a set $A \subset \mathbb{R}^N$. Let $\beta \geq 0$ such that μ satisfies (UAR). Then*

$$\mathcal{H}^\beta(A) > 0. \tag{10.17}$$

In particular,

$$d_c(\mu) \geq \beta . \tag{10.18}$$

Proof. Let $(X_i)_{i \in I}$ be any countable covering of A. Every X_i is contained in a ball B_i with a radius equal to diam(X_i). So, by (UAR),

$$1 = \mu(\mathbb{R}^N) = \mu(A) \leq \sum_{i \in I} \mu(B_i) \leq C \sum_{i \in I} \mathrm{diam}(X_i)^\beta ,$$

from which we get

$$\sum_{i \in I} \mathrm{diam}(X_i)^\beta \geq C^{-1} > 0 .$$

Corollary 10.22. *Let μ be an Ahlfors regular probability measure in dimension $\beta \geq 1$. By (10.16) and (10.18) and since $\beta = \max\{\beta, 1\} \geq \max\{d_M(\mu), 1\}$, the lower and upper bounds stated in Remark 10.13 and Theorem 10.14 for $d_s(\mu)$ and $d(\mu)$, respectively, give*

$$d_c(\mu) = d_s(\mu) = d(\mu) = d_M(\mu) = \beta .$$

This guarantees that, in the case of an Ahlfors regular probability measure, all the geometrical dimensions $d_c(\mu)$, $d_s(\mu)$ and $d_M(\mu)$, and the irrigability dimension $d(\mu)$ are equal to the Ahlfors dimension β.

Corollary 10.23. *An Ahlfors regular probability measure μ of dimension $\beta \geq 1$, is α-irrigable for all $\alpha \in]0, 1[$ s.t. $d_\alpha > \beta$, i.e., for all $\alpha \in]\frac{1}{\beta'}, 1[$, and is not irrigable for all $\alpha \in]0, 1[$ s.t. $d_\alpha \leq \beta$, i.e., for all $\alpha \in]0, \frac{1}{\beta'}]$.*

Proof. Let $\alpha \in]0, 1[$. If $d_\alpha \neq \beta = d(\mu)$, the thesis follows from Remark 10.6. Moreover, when $d_\alpha = \beta$, by Theorem 10.15 it is clear that an Ahlfors regular probability measure of dimension $\beta = d_\alpha$ is not α-irrigable. Indeed, by Lemma 10.21, it cannot be concentrated on a d_α-negligible set.

We shall make use of this last argument when in Section 10.4 we shall show that, in general, $d(\mu) = \inf\{d_\alpha : \mu \text{ is } \alpha\text{-irrigable}\}$ is not a minimum, see Example 10.33.

10.2 Lower Bound on $d(\mu)$

This section is devoted to the proof of Theorem 10.15 from which $d_c(\mu) \leq d(\mu)$ trivially follows.

Lemma 10.24. *Let χ be a pattern in $\mathbb{R}^N$ with finite energy and $r > 0$, then*

$$\mu_\chi(\mathbb{R}^N \setminus B_r(S)) \leq \left(\frac{E^\alpha(\chi)}{r}\right)^{\frac{1}{\alpha}} . \tag{10.19}$$

In particular, if $r \geq (E^\alpha(\chi))^{1-\alpha}$, then

$$\mu_\chi(\mathbb{R}^N \setminus B_r(S)) \leq E^\alpha(\chi) . \tag{10.20}$$

Proof. Taking into account that the less expensive way to carry some part of the fluid out of $B_r(S)$ is to move it in a unique tube in the radial direction and to leave the other part in the source point, we have

$$[\mu_\chi(\mathbb{R}^N \setminus B_r(S))]^\alpha r \leq E^\alpha(\chi) ,$$

from which the thesis follows.

Lemma 10.25. *Let χ be an optimal pattern for E^α. Then for every $\varepsilon > 0$ there is $A \subset \mathbb{R}^N$ such that*

1. A can be covered by finitely many balls $B_i = B_{r_i}(x_i)$, s.t. $\sum_i (r_i)^{d_\alpha} < \varepsilon$;
2. $\mu_\chi(\mathbb{R}^N \setminus A) \leq \varepsilon$.

Proof. By Theorem 7.12 we can find a finite number of points $\{x_i\}_{i=1}^k$ in $\mathbb{R}^N$ such that, by denoting by χ_i the subtree of χ starting from x_i and by $\varepsilon_i = E^\alpha(\chi_i)$,

$$\sum_{i=1}^k \varepsilon_i < \varepsilon \quad \text{and} \quad (\mu_\chi - \sum_{i=1}^n \mu_{\chi_i})(X) < \varepsilon . \tag{10.21}$$

Taking $r_i = (E^\alpha(\chi_i))^{1-\alpha} = (\varepsilon_i)^{1-\alpha}$, $i \in \{1, \ldots, k\}$, yields

$$\sum_i r_i^{d_\alpha} = \sum_i \varepsilon_i < \varepsilon .$$

From (10.20) in Lemma 10.24 we also have

$$\mu_{\chi_i}(\mathbb{R}^N \setminus B_{r_i}(x_i)) \le \varepsilon_i .$$

Applying (10.21), $A = \bigcup_{i=1}^k B_{r_i}(x_i)$ satisfies

$$\mu_\chi(\mathbb{R}^N \setminus A) \le (\mu_\chi - \sum_{i=1}^k \mu_{\chi_i})(\mathbb{R}^N) + \sum_{i=1}^k \mu_{\chi_i}(\mathbb{R}^N \setminus B_{r_i}(x_i)) < \varepsilon + \sum_{i=1}^k \varepsilon_i < 2\varepsilon .$$

Proof. (of Theorem 10.15) By assumption there exists a pattern χ with finite cost irrigating μ. We can take χ optimal. Thus for every $n \in \mathbb{N}$, we can apply Lemma 10.25 to χ with $\varepsilon = 2^{-n} > 0$. Therefore for all $n \in \mathbb{N}$ there exists $A_n \subset \mathbb{R}^N$ which satisfies 1) and 2) of Lemma 10.25 for $\varepsilon = 2^{-n}$. For a fixed $h \in \mathbb{N}$ let us denote by $D_h = \bigcap_{n>h} A_n$. Then

$$\mu_\chi(\mathbb{R}^N \setminus D_h) = \mu_\chi(\bigcup_{n>h} \mathbb{R}^N \setminus A_n) \le \sum_{n>h} \mu_\chi(\mathbb{R}^N \setminus A_n) \le \sum_{n>h} \frac{1}{2^n} = \frac{1}{2^h} .$$
$$(10.22)$$

Since $D_h \subset A_n$ for all $n > h$, by Lemma 10.25(1), D_h is covered by a finite number k of balls with radius r_i satisfying $\sum_{i=1}^k r_i^{d_\alpha} < 2^{-n}$. From the definition of Hausdorff outer measure it follows that $\mathcal{H}^{d_\alpha}(D_h) = 0$. For all $i \in \mathbb{N}$

$$\mu_\chi(\mathbb{R}^N \setminus \bigcup_{h \in \mathbb{N}} D_h) \le \mu_\chi(\mathbb{R}^N \setminus D_i) \le \frac{1}{2^i} ,$$

therefore we have

$$\mu_\chi(\mathbb{R}^N \setminus \bigcup_{h \in \mathbb{N}} D_h) = 0$$

and so μ is concentrated on $\bigcup_{h \in \mathbb{N}} D_h$. Since, for all $h \in \mathbb{N}$, $\mathcal{H}^{d_\alpha}(D_h) = 0$, μ is concentrated on a d_α-negligible set.

Proof. (of the lower bound $d_c(\mu) \le d(\mu)$ in Theorem 10.14) By Theorem 10.15 we have proved in particular that for every $\alpha \in]0, 1[$, if μ is α-irrigable then $d_c(\mu) \le d_\alpha$. By the definition of $d(\mu)$, taking the infimum on d_α in the above inequality, the thesis follows.

10.3 Upper Bound on $d(\mu)$

The main goal of this section is the proof of the following theorem, from which the upper bound on $d(\mu)$ stated in Theorem 10.14 easily follows.

Theorem 10.26. *Let μ be a probability measure and $\alpha \in]0, 1[$, then μ is α-irrigable provided $d_M(\mu) < d_\alpha$.*

Proof. If $d_M(\mu) < d_\alpha$, we can fix a constant β such that $d_M(\mu) < \beta < d_\alpha$.

Given $n \in \mathbb{N}$, $n \geq 1$, let us consider a covering of $\mathrm{supp}(\mu)$ consisting of balls with radius 2^{-n}. Let us call P_n the set made of the centers of such balls and let us set $P_0 = \{S\}$. We introduce for $n \geq 1$ the map $\gamma_n : P_n \to P_{n-1}$ which chooses, for every point $x \in P_n$, one of the closest points $\gamma_n(x) \in P_{n-1}$. It is easy to see that for $n \geq 2$ (and for $n \geq 1$, with a suitable choice of S and a normalization of the diameter of the support of μ)

$$\forall x \in P_n, \quad |x - \gamma_n(x)| \leq 3 \cdot 2^{-n} . \tag{10.23}$$

Moreover, by Lemma 10.11, since $d_M(\mu) < \beta$, we can choose P_n and a constant $C > 0$ so that

$$\mathrm{card}(P_n) \leq C \left(2^{-n}\right)^{-\beta} = C \, 2^{n\beta} . \tag{10.24}$$

Let us now put a total order on P_n. On each center $x \in P_n$ we shall put the mass

$$m_x^n = \mu(B_{2^{-n}}(x) \setminus \bigcup_{y < x} B_{2^{-n}}(y)) .$$

In this way we get a probability measure $\mu_n = \sum_{x \in P_n} m_x^n \delta_x$ such that $\mu_n \rightharpoonup \mu$. By (10.24) we have:

$$\forall i \in \{1, \ldots, n\}, \quad k_i = \mathrm{card}(P_i) \leq C \, 2^{i\beta} , \tag{10.25}$$

while, by (10.23),

$$\forall i \in \{1, \ldots, n\}, \quad l_i = \max_{x \in P_i} |x - \gamma_i(x)| \leq 3 \cdot 2^{-i} . \tag{10.26}$$

As in Definition 5.7, let us denote by $\boldsymbol{P}_n$ the pattern associated with the hierarchy of collectors $(P_i, \gamma_i)_{1 \leq i \leq n}$ which irrigates μ_n. We obtain from Corollary 5.9, (10.25), and (10.26)

$$E^\alpha(\boldsymbol{P}_n) \leq \sum_{i=1}^n (k_i)^{1-\alpha} l_i \leq 3C^{1-\alpha} \sum_{i=1}^n 2^{-i(-\beta(1-\alpha)+1)}$$

$$= 3C^{1-\alpha} \sum_{j=1}^n 2^{-jb} \leq \frac{3C^{1-\alpha}}{2^b - 1} ,$$

where since $\beta < d_\alpha$,

$$b = -\beta(1 - \alpha) + 1 > 0 .$$

The independence on n of the above bound allows us to build a sequence of traffic plans $(\boldsymbol{P}_n)_{n \in \mathbb{N}}$ to which we can apply the compactness Theorem 3.28. The limit traffic plan $\boldsymbol{P}$ has finite energy and is such that $\mu_{\boldsymbol{P}} = \mu$. By Proposition 7.4, it is a pattern.

Proof. of Theorem 10.14 (upper bound $d(\mu) \leq \max\{d_M(\mu), 1\}$) Arguing by contradiction, let us suppose $d(\mu) > \max\{d_M(\mu), 1\}$. Then there exists a constant $\alpha \in]0, 1[$ such that $d_M(\mu) < d_\alpha < d(\mu)$. From one side $d_M(\mu) < d_\alpha$, so we have from Theorem 10.26 that μ is α-irrigable. On the other side $d_\alpha < d(\mu)$, so we get from Remark 10.6 (2) that μ cannot be α-irrigable.

10.4 Remarks and Examples

Lemma 10.27. *Let $\alpha \in]0, 1[$, ν and μ be two positive measures on $\mathbb{R}^N$ such that $\nu = g\mu$ for some $g \in L^\infty(d\mu)$, $g \geq 0$. If μ is α-irrigable then also ν is α-irrigable and $E^\alpha(\nu) \leq \|g\|_\infty^\alpha E^\alpha(\mu)$.*

Proof. Let $\boldsymbol{P}$ be an optimal traffic plan (and therefore a pattern) irrigating μ. Set $\boldsymbol{P}_\nu = (g \circ \pi_\infty) \cdot \boldsymbol{P}$. Then $\mu^+(\boldsymbol{P}) = \pi_{\infty\sharp}\boldsymbol{P} = \mu$ and

$$\mu^+(\boldsymbol{P}_\nu) = \pi_{\infty\sharp}((g \circ \pi_\infty) \cdot \boldsymbol{P}) = g \cdot \mu = \nu.$$

Moreover, since $|x|_{\boldsymbol{P}_\nu} \leq \|g\|_\infty |x|_{\boldsymbol{P}}$, by formula (4.5) we have that $E^\alpha(\boldsymbol{P}_\nu) \leq \|g\|_\infty^\alpha E^\alpha(\boldsymbol{P})$ and the last conclusion follows.

The simple idea that the irrigability of a probability measure μ depends only on the dimension of the support is false. Indeed, $d_s(\mu)$ and $d(\mu)$ are not comparable in general, even if the dimensions $d_c(\mu)$ and $d_M(\mu)$ which respectively give a lower and an upper bound on $d_s(\mu)$ are also bounds for $d(\mu)$, as stated in Remark 10.13 and Theorem 10.14.

It is easy to see that, in general, $d_s(\mu) \not\leq d(\mu)$, as stated in the following example.

Example 10.28. There exist probability measures μ such that $d_s(\mu) = N$ (maximum possible value) and $d(\mu) = 1$ (minimum possible value) i.e. which are α-irrigable for all $\alpha \in]0, 1[$.

Proof. Let us call B the unit ball of $\mathbb{R}^N$, $S = 0$ and let $\tilde{B} = \{x_i : i \in \mathbb{N}\}$ be the countable set consisting in the points of B with rational coordinates.

Let us consider $\mu = \sum_{n\geq 1} \left(\frac{1}{2}\right)^n \delta_n$ where, $\forall n \in \mathbb{N}$, δ_n is a Dirac mass centered in x_n. By construction, $d_s(\mu) = N$. We shall prove that μ is α- irrigable for all $\alpha \in]0, 1[$, i.e. $d(\mu) = 1$. Let χ be the pattern which carries the mass $\frac{1}{2^n}$ at unitary speed carries from S to the n-th point of $\tilde{B}$. Then for any $\alpha \in]0, 1[$ we have $E^\alpha(\chi) \leq \sum_{n\geq 1} \left(\frac{1}{2}\right)^{\alpha n} = \frac{1}{2^\alpha - 1} < +\infty$. Since by construction $\mu_\chi = \mu$, μ is α-irrigable.

Definition 10.29. *We shall say that a measure μ is a* discrete measure *if*

$$\mathrm{card}(\mathrm{supp}(\mu)) < \infty$$

and we shall call $\mathrm{card}(\mathrm{supp}(\mu))$ *the "resolution" of μ.*

In order to show that also the converse inequality is, in general, not true, we shall point out the following property.

Proposition 10.30. *Let $\alpha \in]0, 1[$ and μ be a probability measure which is not α-irrigable. Then for all $n \in \mathbb{N}$ it is possible to find a discrete approximation $\tilde{\mu}$ of μ, of sufficiently high resolution (see Definition 10.29), such that any pattern $\tilde{\chi}$ which irrigates $\tilde{\mu}$ has a cost $E^\alpha(\tilde{\chi}) \geq n$.*

Proof. Assume by contradiction that we can find a sequence $(\tilde{\mu}_n)_{n \in \mathbb{N}}$ of discrete approximations of μ weakly converging to μ and a sequence $(\tilde{\chi}_n)_{n \in \mathbb{N}}$ of patterns irrigating $\tilde{\mu}_n$ such that $\forall n \in \mathbb{N}$, $E^\alpha(\tilde{\chi}_n) < c$. Then we could apply the compactness Theorem 3.28 and Proposition 7.4 to obtain a pattern $\overline{\chi}$ with finite cost irrigating μ. $\quad\square$

Example 10.31. There exists a probability measure μ with a countable support which is not α-irrigable for $\alpha = \frac{1}{N'}$.

Proof. Let B be the unit ball of $\mathbb{R}^N$, let μ_L be the normalized Lebesgue measure on B and $\alpha = \frac{1}{N'}$. Since by Theorem 10.15 μ_L is not α-irrigable, by Proposition 10.30 we can consider a discretization μ_1 of μ_L such that for any pattern $\tilde{\chi}_1$ which irrigates μ_1, $E^\alpha(\chi_1) \geq 1$. Analogously, let μ_2 be a discretization of $\frac{1}{2}\mu_L$ distributed on $\frac{1}{2}B$ such that for any pattern χ_2 which irrigates μ_2, $E^\alpha(\chi_2) \geq 2$. Recursively, for any $n \in \mathbb{N}$ let μ_n be a discretization of $\frac{1}{2^n}\mu_L$ restricted to $\frac{1}{n}B$ such that for any pattern χ_n which irrigates it, we have

$$E^\alpha(\chi_n) \geq n. \tag{10.27}$$

Let $\mu = \sum_{n \geq 1} \mu_n$ and let us remark that $\mathrm{supp}(\mu) = \bigcup_{n \geq 1} \mathrm{supp}(\mu_n) \cup \{0\}$ and therefore, since $\mathrm{supp}(\mu_n)$ is a finite set for all $n \geq 1$, $\mathrm{supp}(\mu)$ is countable.

Let us show that μ is not α-irrigable. Indeed, the α-irrigability of μ would imply, by Lemma 10.27 (since $\mu_n \leq \mu$ for all $n \geq 1$), that any μ_n is α-irrigable with a bounded cost and this is in contradiction with (10.27). $\quad\square$

Using the measure μ of Example 10.31, we have the following statement.

Example 10.32. There exist probability measures μ such that $d_s(\mu) = 0$ (minimum possible value) and $d(\mu) = N$ (maximum possible value).

Notice that this implies that, in general, $d(\mu) \not\leq d_s(\mu)$.

We have stated in Section 10.1 that the information that, for a probability measure μ and a real number $\alpha \in]0, 1[$, the critical dimension d_α coincides with the irrigability dimension $d(\mu)$ (i.e. $\alpha = \frac{1}{(d(\mu))'}$), does not allow to decide whether the measure is irrigable or not. Examples 10.33 and 10.35 will motivate this claim.

Example 10.33. Let μ be an Ahlfors regular probability measure in dimension $\beta \geq 0$. Then μ is not α-irrigable if $d_\alpha = \beta = d(\mu)$.

Proof. The thesis follows from Corollary 10.23.

Remark 10.34. One has Ahlfors regular measures for every dimension $\beta < N$. Indeed, let $\mathcal{C}$ be a selfsimilar (Cantor) set of $\mathbb{R}^N$ with dimension $\beta > 0$. Let us call $\mathcal{H}^\beta_{\lfloor \mathcal{C}}$ the Hausdorff measure *distributed* on $\mathcal{C}$, i.e. the measure on $\mathbb{R}^N$ defined setting $\forall X \subset \mathbb{R}^N$

$$\mathcal{H}^\beta_{\lfloor \mathcal{C}}(X) = \mathcal{H}^\beta(X \cap \mathcal{C}) .$$

Then $\mathcal{H}^\beta_{\lfloor \mathcal{C}}$ is Ahlfors regular with dimension β.

Example 10.35. There exist some measures μ for which $d(\mu)$ is a minimum, i.e. there exist some measures μ and some exponents $\alpha \in]0, 1[$ such that $d(\mu) = d_\alpha$ and μ is α-irrigable.

Proof. Indeed, let us fix $\alpha \in]0, 1[$ and let $(\mathcal{C}_n)_{n\in\mathbb{N}}$ be a sequence of self similar (Cantor) sets in $\mathbb{R}^N$ with dimension d_{α_n} where $(\alpha_n)_{n\in\mathbb{N}}$ is a sequence converging to α from below. By Corollary 10.22 and Remark 10.34 we know that $d(\mathcal{H}^{d_{\alpha_n}}_{\lfloor \mathcal{C}_n}) = d_{\alpha_n} < d_\alpha$ and, by Remark 10.6 (1), we get that $\mathcal{H}^{d_{\alpha_n}}_{\lfloor \mathcal{C}_n}$ is α-irrigable. Let us consider a suitable sequence $(\varepsilon_n)_{n\in\mathbb{N}}$ of positive real numbers, sufficiently small to allow us to consider $\mu = \sum_{n\in\mathbb{N}} \varepsilon_n \mathcal{H}^{d_{\alpha_n}}_{\lfloor \mathcal{C}_n}$. We know, by Lemma 10.27 that also $\varepsilon_n \mathcal{H}^{d_{\alpha_n}}_{\lfloor \mathcal{C}_n}$ are irrigable and we call χ_n an optimal pattern irrigating $\varepsilon_n \mathcal{H}^{d_{\alpha_n}}_{\lfloor \mathcal{C}_n}$ from a fixed source S. Under the choice of a sufficiently infinitesimal sequence of coefficients $(\varepsilon_n)_{n\in\mathbb{N}}$, we have $\sum_{n\in\mathbb{N}} E^\alpha(\chi_n) < +\infty$.

Now let us consider the union χ of the sequence of patterns χ_n (see Definition 5.3) so that by Lemma 5.2(*ii*), we have

$$\mu_\chi = \mu \tag{10.28}$$

and

$$E^\alpha(\chi) \le \sum_{n\in\mathbb{N}} E^\alpha(\chi_n) < +\infty . \tag{10.29}$$

Taking into account (10.28), (10.29) and replacing χ by the optimal traffic plan irrigating μ, which by Proposition 7.4 is a pattern, we obtain the α-irrigability of μ and therefore $d(\mu) \le d_\alpha$. Moreover $d(\mu) \ge d_\alpha$. Indeed since, for all $n \in \mathbb{N}$, $\mu \ge \varepsilon_n \mathcal{H}^{d_{\alpha_n}}_{\lfloor \mathcal{C}_n}$ we get by Lemma 10.27 and Remark 10.34 that $d(\mu) \ge d(\varepsilon_n \mathcal{H}^{d_{\alpha_n}}_{\lfloor \mathcal{C}_n}) = d(\mathcal{H}^{d_{\alpha_n}}_{\lfloor \mathcal{C}_n}) = d_{\alpha_n} \to d_\alpha$.

Remark 10.36. A complementary analysis of the irrigability of pairs of Borel measures can be found in [69]. Besides irrigability, two complementary concepts are introduced. We say that a nonnegative Borel measure μ with compact support in $\mathbb{R}^N$ is purely α-non-irrigable, if any positive Borel measure $\mu' \le \mu$ is not α-irrigable. We say that the measure μ is marginally α-non-irrigable if μ is not α-irrigable and there is not a purely α-non-irrigable positive measure $\mu' \le \mu$, in other words, if given any positive measure $\mu' \le \mu$, there is a further positive measure $\mu'' \leq \mu'$ which is α-irrigable. The authors observe that Ahlfors-regular measures in dimension $\beta \geq 1$ are purely α-non-irrigable if $\beta > \left(\frac{1}{\alpha}\right)'$. Then the authors prove that any nonnegative Borel measure with finite mass can be decomposed as the sum of two measures, a purely α-non-irrigable one and either an α-irrigable one or a marginally α-non-irrigable, $\alpha \in [0, 1]$. The latter ones can be written as a countable sum of α-irrigable measures. They also prove that if two positive Borel measures with equal total finite mass and compact support can be transported at finite E^α cost, then their purely α-non-irrigable components must be equal and the complementary parts can be irrigated at finite cost. This implies, in particular, that if the two given measures are concentrated on disjoint sets and can be connected at finite cost, then they are both either α-irrigable or marginally α-non-irrigable. Moreover, if the distributional supports are disjoint, they are both α-irrigable.

11 The Landscape of an Optimal Pattern

Consider the irrigation of an arbitrary measure on a domain X starting from a single source $\mu^+ = \delta_0$. In this chapter we define following Santambrogio [75] a function $z(x)$ that represents the elevation of the landscape associated with an optimal traffic plan. This function was known in the geophysical community under a very strong discretization [6], [73]. The continuous landscape function will be proven to share all the properties that hold in the discrete case, in particular the fact that, at a point x_0 of the irrigation network, $z(x)$ has maximal slope in the direction of the network itself and that this slope is given by a power of the multiplicity, $|x|_P^{\alpha-1}$. Section 11.1 describes the physical discrete model of joint landscape-river network evolution. In Section 11.2 we will prove a general first variation inequality for the energy E^α when μ^+ varies. Section 11.3 proves that the definition of $z(x)$, which will be defined as a path integral along the network from the source to x, does not depend on the path but only on x. Section 11.4 is devoted to a general semicontinuity property of $z(x)$ and Section 11.5 to its Hölder continuity when the irrigated measure dominates Lebesgue on X. The whole chapter follows closely the seminal Santambrogio paper [75].

11.1 Introduction

11.1.1 Landscape Equilibrium and OCNs in Geophysics

Geophysics yields problems very similar to the Gilbert-Steiner problem, in the study of river basins. A quite comprehensive reference is [73]. The specific subject dealt with in this chapter is developed both in [73] and in [6] as well. While studying the configuration of a river basin, the main objects are two: the landscape elevation, which is a function z giving the altitude of any point of the region we are considering, and a river network P, which is the datum of all the streams that concur to bring water (which falls on the region as rain) to a single point (where a lake is supposed to be present). This is why we denote this river network directly as a traffic plan. A first link between landscape and river network is the fact that at any point the direction followed by water is the direction of steepest descent of z. Thus, the support of P can be drawn by following all steepest descent lines of z. Hence, once we know z we are able to deduce P and also to compute the multiplicity $\theta(x) = |x|_P$ at any point x, that is the quantity of water passing through x while following the steepest descent lines of z. In geophysical terms $\theta(x)$ is the area of the basin of x. At first the interest is towards an evolution model, which allows z and $\boldsymbol{P}$ (and hence θ) to depend on time as well. The evolution of z is ruled by an erosion equation of the form

$$\frac{\partial z}{\partial t} = -\theta|\nabla z|^2 + c, \qquad (11.1)$$

where ∇z is the spatial gradient of z and c is a positive constant. The idea is that the erosion effect increases both with the quantity of water and with the slope. The constant c is called uplift and takes care of the fact that all the material brought down by erosion in the end is uniformly redistributed from below in the whole region as a geomorphological effect. Equation (11.1) is in fact a simplified version of other more general evolution equations involving higher order terms. The following phenomenon concerning solutions of (11.1) can be empirically observed: approximately, up to a certain time scale both z and θ (i.e. $\boldsymbol{P}$) move, in a very strong erosional evolution; then, up to a larger time scale the network is almost constant, letting $\theta(x,t) = \theta(x)$ depend on the position only, and the landscape function evolves without changing its lines of maximal slope; finally there is a much larger time scale such that z approximatively agrees with a landscape equilibrium, i.e. a stationary solution of (11.1). We are interested in studying landscape equilibria. In this case the steepest descent condition, that we can read as "∇z follows the direction of the network", is completed by a second one which we get by imposing $\partial z/\partial t = 0$ in (11.1). This leads to $|\nabla z| = c^{1/2}\theta^{-1/2}$ and this last condition is called *slope-discharge relation* . It is explicitly suggested in [6] that in (11.1) one could change the exponents of θ and $|\nabla z|$ (preserving anyway the increasing behavior with respect to both variables), thus obtaining different slope-discharge relationships. In general $|\nabla z| = c\,\theta^{\alpha-1}$ and the physically interesting case is when the exponent α is very close to $1/2$.

To find landscape equilibria in a discrete setting, [6] uses a regular square grid. Functions defined on the pixels of the grid and vanishing at a given point x_0 representing the outlet are considered, as well as networks composed by edges of the grid, directed from every point to one of the neighbors. The above mentioned ideas lead to the following algorithm.

Joint Landscape-Network Evolution

- As we already mentioned, the conditions on the direction of the water allow to reconstruct a network $\boldsymbol{P}$ from a function. In fact, given a function z with no local minima other than x_0, one can always follow the maximal slope paths of z.
- These are obtained by linking any point x of the grid to a point which realizes the minimum of z among the neighbors of z. Notice in particular that

these paths are only composed by edges following the two main directions of the grid.

- In this way a network $\boldsymbol{P} = \boldsymbol{P}(z)$ can be deduced from z.
- On the other hand, the slope-discharge condition allows to reconstruct a function from a network $\boldsymbol{P}$, provided it is tree-shaped.
- In order to make this reconstruction, first compute the multiplicities of the points of the network: at a point x its multiplicity $\theta(x)$ is the number of points which find x on their way to the outlet (this works under the assumption that the quantity of rain falling down at any pixel is the same, i.e. rain falls uniformly on the grid). See also Figure 11.1, where the multiplicity of a point x_i is computed as the number of points in the area A_i.
- Then set $z(x_0) = 0$ and for any other point x consider the only path on $\boldsymbol{P}$ linking x_0 to x. Set $z(x) = \sum_i \theta(x_i)^{\alpha-1}$, where the x_i's are the points on the path. In Figure 11.1 the path linking x_0 to x is shown.
- In this way we get a function $z = z(\boldsymbol{P})$.

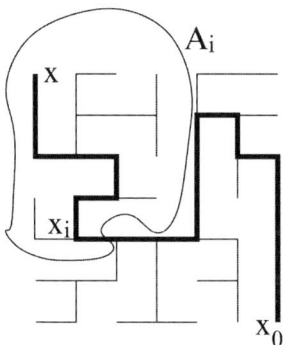

Fig. 11.1. The path from x to x_0 and the multiplicity of x_i.

In general it will not be true that a function $z(\boldsymbol{P})$ has maximal slope in the direction of the network $\boldsymbol{P}$. Finding a landscape equilibrium means exactly satisfying both conditions at a time, through a fixed point problem. The algorithm starts from a tree-shaped network $\boldsymbol{P}$, builds the function $z(\boldsymbol{P})$, and then the new network $\boldsymbol{P}' = \boldsymbol{P}(z(\boldsymbol{P}))$. If $\boldsymbol{P}' = \boldsymbol{P}$, the landscape function $z = z(\boldsymbol{P})$ is a landscape equilibrium.

The important idea presented in [6] is the relation between landscape equilibria and Optimal Channel Networks (OCNs in literature, see for instance [74], [72] and [48]). An OCN is a network $\boldsymbol{P}$ minimizing a certain dissipated energy. The dissipated energy in a system satisfying the slope-discharge relation is the total potential energy that water loses on the network. For each pixel we have a quantity of water θ which falls down towards the next pixel and its elevation decreases by a quantity which is proportional to $|\nabla z|$ and hence to $\theta^{\alpha-1}$. Hence, the total energy loss is given by

$$E(z) = \sum_i \theta(x_i)\theta(x_i)^{\alpha-1} = \sum_i \theta(x_i)^{\alpha}. \qquad (11.2)$$

It is clear that this energy is the Gilbert energy (no length of segments is involved because in a regular grid they all have the same, given, length). What is proven in [6] is the following theorem.

Theorem 11.1. *If P is an OCN minimizing the energy (11.2), then the landscape function $z = z(P)$ reconstructed from it is in fact an equilibrium. This actually means that not only the slope of z in the direction of the network is given by $\theta^{\alpha-1}$, which is true by construction, but also that this direction is the direction of maximal slope.*

Our first aim will be to get a continuous version of this result. Notice that the problems studied in [6] and in the other papers on the subject have undergone a very strong discretization. In fact, they work with a discretization μ on a regular grid of the Lebesgue measure and $\nu = \delta_{x_0}$ and with the extra constraint that only grid edges e_h are allowed. Compared to continuous models there is a loss of rotational invariance, a fixed scale effect due to the mesh. Thus, several questions concerning the river basin may lose their meaning (for instance questions about the interfaces between two separated parts of the basin and points where the water takes two different directions, and most regularity issues). Thanks to the continuous version we will be able to prove uniform regularity properties for the landscape.

11.2 A General Development Formula

In this section we will develop in a useful way the variation of the functional E^{α} when passing from a traffic plan P to a traffic plan P'. The formulation (4.5) of the traffic plan energy $E^{\alpha}(P) = \int_{\mathbb{R}^N} |x|_X^{\alpha} d\mathcal{H}^1(x)$ will be crucial. Let us denote by K_{arc} the set of those curves in K which are parameterized by arc length and by K_{inj} the set of curves in K which are injective on $[0, T(\gamma)[$. In the sequel we will often identify a curve with its image, in the sense that sometimes we will write γ instead of $\gamma([0, T(\gamma)]) = \gamma([0, +\infty[)$. **In this chapter, we assume that the traffic plan associated to P is parameterized by arc length.** For every $\gamma \in K$ define

$$Z_P(\gamma) = \int_0^{T(\gamma)} |\gamma(t)|_P^{\alpha-1} dt,$$

so that $E^{\alpha}(P) = \int_K Z_P \, dP$.

Theorem 11.2. *Let P and P' be probability measures on K and $\Delta P = P' - P$. Let us suppose that both P and ΔP are concentrated on $K_{arc} \cap K_{inj}$ and $\int_K Z_P \, d|\Delta P| < +\infty$. Then*

$$E^\alpha(P') \le E^\alpha(P) + \alpha \int_K Z_P \, d\Delta P - \alpha(1-\alpha) \int_{\mathbb{R}^d} |x|^2_{\Delta P} \, d\mathcal{H}^1(x), \quad (11.3)$$

where $|x|_{\Delta P} = |x|_{P'} - |x|_P$.

Proof. Let us first prove that $\mathcal{H}^1(S_{P'} \setminus S_P) = 0$. For any point $x \in S_{P'} \setminus S_P$ $|x|_P = 0$ and $|x|_{\Delta P} > 0$ necessarily hold. Hence it is sufficient to prove that the integral of $|x|_{\Delta P}$ on this set vanishes to get the desired result. Yet we have

$$\int_{S_{P'} \setminus S_P} |x|_{\Delta P} d\mathcal{H}^1(x) = \int_{S_{P'} \setminus S_P} d\mathcal{H}^1(x) \int_K d\Delta P(\gamma) \mathbb{1}_{x \in \gamma}$$

$$= \int_K d\Delta P(\gamma) \mathcal{H}^1(\gamma \cap (S_{P'} \setminus S_P)).$$

The second assumption of the theorem implies that for ΔP−a.e. curve γ the quantity $Z_P(\gamma)$ is finite and hence, for a.e. t, we have $\gamma(t) \in S_P$. Since γ is 1−Lipschitz continuous, this yields $\mathcal{H}^1(\gamma \setminus S_P) = 0$. Hence we have $\int_{S_{P'} \setminus S_P} |x|_{\Delta P} d\mathcal{H}^1(x) = 0$, which proves $\mathcal{H}^1(S_{P'} \setminus S_P) = 0$.

Since both P and P' are concentrated on $K_{arc} \cap K_{inj}$, to evaluate the energy we can use the expression in (4.5) and get

$$E^\alpha(P') = \int_{S_P} (|x|_P + |x|_{\Delta P})^\alpha \, d\mathcal{H}^1(x)$$

$$\le E^\alpha(P) + \alpha \int_{S_P} |x|^{\alpha-1}_P |x|_{\Delta P} d\mathcal{H}^1(x) - \alpha(1-\alpha) \int_{S_P} |x|^2_{\Delta P} d\mathcal{H}^1(x), \quad (11.4)$$

where we have used the fact that $S_{P'} \subset S_P$ up to $\mathcal{H}^1$−negligible sets and the concavity inequalities

$$(t+s)^\alpha \le t^\alpha + \alpha t^{\alpha-1} s - \alpha(1-\alpha)(\max\{t, t+s\})^{\alpha-2} s^2 \le t^\alpha + \alpha t^{\alpha-1} s - \alpha(1-\alpha) s^2$$

(this last inequality being valid when both t and $t+s$ belong to $]0, 1]$). Let us now work on the second term of the last sum we obtained. We have

$$\int_{S_P} |x|^{\alpha-1}_P |x|_{\Delta P} d\mathcal{H}^1(x) = \int_{S_P} d\mathcal{H}^1(x) \int_K d\Delta P(\gamma) |x|^{\alpha-1}_P \mathbb{1}_{x \in \gamma}.$$

Here we want to change the order of integration and to do this we check what happens in absolute value:

$$\int_{S_P} d\mathcal{H}^1(x) \int_K d|\Delta P|(\gamma) |x|^{\alpha-1}_P \mathbb{1}_{x \in \gamma} \quad (11.5)$$

$$= \int_K d|\Delta P|(\gamma) \int_0^{T(\gamma)} |\gamma(t)|^{\alpha-1}_P dt = \int_K Z_P d|\Delta P| < +\infty. \quad (11.6)$$

After changing the integration order, the first equality relies on the fact that $|\Delta P|$−a.e. we have $\mathcal{H}^1(\gamma \setminus S_P) = 0$ and, γ being parametrized by arc length, the $\mathcal{H}^1$−integral on its image may become an integral in dt on $[0, T(\gamma)]$. The finiteness of the last integral in (11.5) allows us to change the order of integration between ΔP and $\mathcal{H}^1$ and by analogous computations we get

$$\int_{S_P} d\mathcal{H}^1(x) \int_K d\Delta P(\gamma) |x|_P^{\alpha-1} \mathbb{1}_{x \in \gamma} = \int_K Z_P d\Delta P.$$

Inserting this last equality in (11.4) gives the thesis.

11.3 Existence of the Landscape Function and Applications

Even when not specifically stated, from now on P will be an optimal pattern irrigating an α−irrigable measure μ.

11.3.1 Well-Definedness of the Landscape Function

For any $t \geq 0$ consider an equivalence relation on K given by "the two curves γ_1 and γ_2 are in relation at time t if they agree on the interval $[0, t]$", and denote the equivalence classes by $[\cdot]_t$, so that

$$[\gamma]_t = \{\tilde{\gamma} : \tilde{\gamma}(s) = \gamma(s) \text{ for any } s \leq t\}.$$

For notational simplicity, let us set $|\gamma|_{t,P} := P([\gamma]_t)$.

Definition 11.3. *Given $P \in \mathcal{P}(K)$, a curve $\gamma \in K$ is said to be P−good if*

$$\int_0^{T(\gamma)} |\gamma|_{t,P}^{\alpha-1} dt < +\infty.$$

Observe that if $\gamma \in K$ is a P−good curve, then γ is parameterized by arc-length and the image of γ is a path contained in the support of the traffic plan associated to P. In particular, $|\gamma|_{t,P} = |\gamma(t)|_P$. The notation introduced serves to condensate all these concepts.

The next lemma is a direct application of Lemma 8.4 to the optimality of a pattern from which the subtree starting at $\gamma(t_0)$ has been severed.

Lemma 11.4. *Let P be an optimal pattern. If γ_0 is a curve such that $|\gamma_0|_{t_0,P} > 0$, set $x_0 = \gamma_0(t_0)$, $A = [\gamma_0]_{t_0}$, $\mu_A = \pi_{\infty\sharp}(\mathbb{1}_A \cdot P)$, $\mu' = \mu - \mu_A + P(A)\delta_{x_0}$, $P' = P - \mathbb{1}_A \cdot P + P(A)\delta_{\bar{\gamma}_0}$, where the curve $\bar{\gamma}_0$ is the curve γ_0 stopped at time t_0. Then P' is an optimal pattern irrigating the measure μ'.*

Notice that the probability $\boldsymbol{P}'$ corresponds to a pattern where we have stopped the fibers that continued after x_0 in the pattern $\boldsymbol{P}$. The mass irrigated by those fibers has been concentrated at x_0.

Theorem 11.5. *If γ_0 and γ_1 are two $\boldsymbol{P}-$good curves sharing the same endpoint $\bar{x}$, then $Z_{\boldsymbol{P}}(\gamma_0) = Z_{\boldsymbol{P}}(\gamma_1)$.*

Proof. If the two curves are identical the thesis is easily obtained. If they are not identical, then they must split at a certain time $\bar{t}$ and at least one of them does not stop exactly at $\bar{t}$. Then we can choose two times t_0 and t_1 with $|\gamma_i|_{t_i,\boldsymbol{P}} > 0$ and $\bar{t} \leq t_i \leq T(\gamma_i)$ for $i = 0,\ 1$. Let us set $x_i = \gamma_i(t_i)$ and $l = |x_1 - x_0|$. Then we use the result and the notations of the previous Lemma and we write

$$E^\alpha(\boldsymbol{P}') = E^\alpha(\delta_0, \mu') \leq E^\alpha(\delta_0, \mu'') + E^\alpha(\mu', \mu''), \tag{11.7}$$

where $\mu'' = \mu - \mu_A + \boldsymbol{P}(A)\delta_{x_1}$. Define $\boldsymbol{P}'' = \boldsymbol{P} - \mathbb{1}_A \cdot \boldsymbol{P} + \boldsymbol{P}(A)\delta_{\bar{\gamma}_1}$, where the curve $\bar{\gamma}_1$ is the curve γ_1 stopped at time t_1. Notice that in $\boldsymbol{P}''$ we have eliminated the fibers that passed trough x_0 in the pattern $\boldsymbol{P}$ and we have redirected them to the curve $\bar{\gamma}_1$. The mass irrigated by this new pattern is $\pi_{\infty\sharp}\boldsymbol{P}'' = \mu''$. Then

$$E^\alpha(\delta_0, \mu'') \leq E^\alpha(\boldsymbol{P}'') \leq E^\alpha(\boldsymbol{P}') + \alpha \int_K Z_{\boldsymbol{P}}\, d(\boldsymbol{P}'' - \boldsymbol{P}')$$
$$= E^\alpha(\boldsymbol{P}') + \alpha \boldsymbol{P}(A) \left(\int_0^{t_1} |\gamma_1(t)|_{\boldsymbol{P}}^{\alpha-1}\, dt - \int_0^{t_0} |\gamma_0(t)|_{\boldsymbol{P}}^{\alpha-1} \right).$$

Here we have used Theorem 11.2 to estimate $E^\alpha(\boldsymbol{P}'')$. Actually by this theorem we should have had $Z_{\boldsymbol{P}'}$ instead of $Z_{\boldsymbol{P}}$. Yet we can interchange $Z_{\boldsymbol{P}'}$ and $Z_{\boldsymbol{P}}$ because we have only replaced the measure $\boldsymbol{P}$ on A by a same amount of mass concentrated on $\bar{\gamma}_0$, and on $\gamma_0 \cup \gamma_1$ this does not affect multiplicities. As far as the second term of the sum in (11.7) is concerned it is easy to see that we have

$$E^\alpha(\mu', \mu'') \leq l\boldsymbol{P}(A)^\alpha.$$

By inserting these estimates in (11.7) we get

$$\int_0^{t_0} |\gamma_0(t)|_{\boldsymbol{P}}^{\alpha-1}\, dt - \int_0^{t_1} |\gamma_1|_{\boldsymbol{P}}^{\alpha-1}\, dt \leq \alpha^{-1} l\boldsymbol{P}(A)^{\alpha-1}.$$

Now estimate the length l by

$$l = |x_0 - x_1| \leq |x_0 - \bar{x}| + |\bar{x} - x_1| \leq (T(\gamma_0) - t_0) + (T(\gamma_1) - t_1)$$
$$\leq \boldsymbol{P}(A)^{1-\alpha} \int_{t_0}^{T(\gamma_0)} |\gamma_0(t)|_{\boldsymbol{P}}^{\alpha-1} dt + \boldsymbol{P}(B)^{1-\alpha} \int_{t_1}^{T(\gamma_1)} |\gamma_1(t)|_{\boldsymbol{P}}^{\alpha-1} dt,$$

where $B = [\gamma_1]_{t_1}$. Hence

$$\int_0^{t_0} |\gamma_0(t)|_{\boldsymbol{P}}^{\alpha-1} dt - \int_0^{t_1} |\gamma_1(t)|_{\boldsymbol{P}}^{\alpha-1} dt$$

$$\leq \alpha^{-1} \int_{t_0}^{T(\gamma_0)} |\gamma_0(t)|_{\boldsymbol{P}}^{\alpha-1} dt + \frac{\boldsymbol{P}(B)^{1-\alpha}}{\boldsymbol{P}(A)^{1-\alpha}} \int_{t_1}^{T(\gamma_1)} |\gamma_1(t)|_{\boldsymbol{P}}^{\alpha-1} dt.$$

Notice that we cannot have $|\gamma_i|_{T(\gamma_i),\boldsymbol{P}} > 0$ for both $i = 0, 1$, by the single path property. So, if $|\gamma_1|_{T(\gamma_1),\boldsymbol{P}} = 0$, once we fix t_0 such that $\boldsymbol{P}(A) > 0$, we can choose t_1 so that $\boldsymbol{P}(B) \leq \boldsymbol{P}(A)$ since $\boldsymbol{P}(B) \to 0$ as $t_1 \to T(\gamma_1)$. Otherwise, if $|\gamma_1|_{T(\gamma_1),\boldsymbol{P}} > 0$, we can choose directly $t_1 = T(\gamma_1)$. In both cases we have

$$\int_0^{t_0} |\gamma_0(t)|_{\boldsymbol{P}}^{\alpha-1} dt - \int_0^{t_1} |\gamma_1(t)|_{\boldsymbol{P}}^{\alpha-1} dt$$

$$\leq \alpha^{-1} \left(\int_{t_0}^{T(\gamma_0)} |\gamma_0(t)|_{\boldsymbol{P}}^{\alpha-1} dt + \int_{t_1}^{T(\gamma_1)} |\gamma_1(t)|_{\boldsymbol{P}}^{\alpha-1} dt \right). \tag{11.8}$$

Then we let t_0 and t_1 tend to $T(\gamma_0)$ and $T(\gamma_1)$, according to the criteria for the choice of t_1 we have used so far, and we get at the limit

$$Z_{\boldsymbol{P}}(\gamma_0) - Z_{\boldsymbol{P}}(\gamma_1) \leq 0,$$

because the integral terms on the right hand side of (11.8) tend to zero as a consequence of the fact that γ_0 and γ_1 are both $\boldsymbol{P}$−good curves. By interchanging the role of γ_0 and γ_1 the thesis is proven.

Corollary 11.6. *If two different $\boldsymbol{P}$−good curves γ_0 and γ_1 meet at a certain point $x = \gamma_0(t_0) = \gamma_1(t_1)$, then $|\gamma_0|_{t_0,\boldsymbol{P}} = |\gamma_1|_{t_1,\boldsymbol{P}} = 0$.*

Proof. If one of the two multiplicities $|\gamma_i|_{t_i,\boldsymbol{P}}$ were positive, by Corollary 7.9, a strict inequality between $Z_{\boldsymbol{P}}(\gamma_0)$ and $Z_{\boldsymbol{P}}(\gamma_1)$ should hold. Yet equality has just been proven and this is a contradiction.

Corollary 11.7. *Any $\boldsymbol{P}$−good curve γ is in fact injective on $[0, T(\gamma)]$.*

Proof. The injectivity on $[0, T(\gamma)[$ is already known. Hence, consider the case $\gamma(T(\gamma)) = \gamma(t)$ for $t < T(\gamma)$. This would imply $|\gamma|_{t,\boldsymbol{P}} > 0$ but this contradicts Corollary 11.6, applied to the pair of good curves γ and $\bar{\gamma}$, which is γ stopped at time t.

The result of Theorem 11.5 allows us to define a function on X by the values of $Z_{\boldsymbol{P}}$.

Definition 11.8. *We define the landscape function associated to the traffic plan $\boldsymbol{P}$ as the function $z_{\boldsymbol{P}}$ given by*

$$z_{\boldsymbol{P}}(x) = \begin{cases} Z_{\boldsymbol{P}}(\gamma) & \text{if } \gamma \text{ is } \boldsymbol{P}-\text{good and } x = \gamma(T(\gamma)); \\ +\infty & \text{if no } \boldsymbol{P}-\text{good curve ends at } x. \end{cases}$$

Remark 11.9. It was in fact possible to prove more easily that $\mu-$a.e. the value of z was well defined (in the sense that if on a non negligible set of points x we had two different values for Z_P we would have had the possibility to strictly improve the value of E^α). Yet, we do not want a function z which is defined a.e. but a pointwise defined value, to deal later with pointwise properties, being also concerned with negligible sets such as S_P.

11.3.2 Variational Applications

Some consequences of the existence of the landscape function are presented here.

Definition 11.10. *For any measure $\mu \in \mathcal{P}(X)$ we set $X_\alpha(\mu) = E^\alpha(\mu, \delta_0)$. Thus a measure μ is $\alpha-$irrigable if and only if $X_\alpha(\mu) < +\infty$.*

Corollary 11.11. *For the functional X_α we have the following representation formula*

$$X_\alpha(\mu) = \int_X z\, d\mu,$$

where $z = z_P$ is the landscape function associated to any optimal pattern P irrigating the measure μ.

Proof. It is sufficient to take the formula $X_\alpha(\mu) = E^\alpha(P) = \int_K Z_P dP$ and use the fact that, by Theorem 11.5, $Z_P(\gamma)$ depends only on $\pi_\infty(\gamma)$. This allows to re-write the integral in terms of the image measure $\pi_{\infty\sharp}P = \mu$ and get the thesis.

Corollary 11.12. *If μ is $\alpha-$irrigable, then any landscape function z is finite $\mu-$a.e.*

Proof. Corollary 11.11 yields $\int z\, d\mu = X_\alpha(\mu) < +\infty$ and from this the result is straightforward.

Remark 11.13. As the word "any" in the previous statement suggests, there is no uniqueness for the landscape function, and there is a landscape function for any optimal pattern.

The next theorem will be useful when looking for continuity properties of the landscape function.

Theorem 11.14. *For a given function g on X, such that $||g||_{L^\infty(\mu)} \le 1$ and such that $\int_X g\, d\mu = 0$, set $\mu_1 = \mu(1 + g)$. Then*

$$X_\alpha(\mu_1) \le X_\alpha(\mu) + \alpha \int_X z(x)g(x)d\mu(x),$$

where the function $z = z_P$ is the landscape function according to an arbitrary optimal pattern P irrigating the measure μ.

Proof. We will consider a variation of P given by $P_1 = (1 + (g \circ \pi_\infty)) \cdot P$. Since $\pi_\sharp P_1 = (1 + g) \cdot \mu$, we have

$$X_\alpha(\mu_1) - X_\alpha(\mu) \leq E^\alpha(P_1) - E^\alpha(P).$$

We want to apply Theorem 11.2 to this situation, with $\Delta P = P_1 - P = (g \circ \pi_\infty) \cdot P$. Since ΔP is absolutely continuous with respect to P with bounded density, it is straightforward that both the conditions required by the theorem (ΔP being concentrated on $K_{arc} \cap K_{inj}$ and Z_P being $|\Delta P|$−integrable) are satisfied, so that one gets

$$E^\alpha(P') \leq E^\alpha(P) + \alpha \int_K Z_P d\Delta P.$$

Now use the fact that Z_P depends only on its terminal point $\pi_\infty(\gamma)$ through $Z_P(\gamma) = z(\pi_\infty(\gamma))$ and get

$$\int_K Z_P d\Delta P = \int_X z \, d(\pi_{\infty\sharp}\Delta P) = \int_X z g d\mu.$$

Putting together all the results yields the thesis.

A simple consequence of this theorem may be expressed in terms of derivatives.

Corollary 11.15. *Set $\mu_\varepsilon = \mu + \varepsilon g \cdot \mu$. Then the following derivative inequality holds:*

$$\limsup_{\varepsilon \to 0^+} \frac{X_\alpha(\mu + \varepsilon g \cdot \mu) - X_\alpha(\mu)}{\varepsilon} \leq \alpha \int_X z(x)g(x)d\mu(x).$$

11.4 Properties of the Landscape Function

11.4.1 Semicontinuity

Lemma 11.16. *Given any $P \in \mathcal{P}(K)$, the function $Z_P : K \to \mathbb{R}$ is lower semi-continuous with respect to pointwise convergence.*

Proof. We know by Lemma 3.25 that $x \mapsto |x|_P$ is upper semi-continuous, and hence $x \mapsto |x|_P^{\alpha-1}$ is l.s.c. Then, to prove $\liminf_n Z_P(\gamma_n) \geq Z_P(\gamma)$, fix a time $t_1 < T(\gamma)$ and use $\liminf T(\gamma_n) \geq T(\gamma)$. Thus eventually $T(\gamma_n) > t_1$ and, by Fatou's Lemma, we get

$$\liminf_n Z_P(\gamma_n) \geq \liminf_n \int_0^{t_1} |\gamma_n(t)|_P^{\alpha-1} dt \geq \int_0^{t_1} |\gamma(t)|_P^{\alpha-1} dt.$$

Passing to the limit as $t_1 \to T(\gamma)$ gives the thesis.

Theorem 11.17. *The landscape function z is lower semi-continuous.*

Proof. Consider a sequence $x_n \to x$ and, correspondingly, some $\boldsymbol{P}$−good curves γ_n such that $\pi_\infty(\gamma_n) = x_n$ and $z(x_n) = Z_{\boldsymbol{P}}(\gamma_n)$. We may assume $\sup_n z(x_n) < +\infty$. Since $T(\gamma_n) \leq Z_{\boldsymbol{P}}(\gamma_n) = z(x_n)$, we also have $\sup_n T(\gamma_n) < +\infty$ and we can extract a subsequence (not relabeled) such that $\gamma_n \to \gamma$ uniformly. It is not difficult to prove that $\pi_\infty(\gamma) = x$. Thus, it is sufficient to use Lemma 11.16 to get $Z_{\boldsymbol{P}}(\gamma) \leq \liminf_n Z_{\boldsymbol{P}}(\gamma_n) = \liminf_n z(x_n)$. This implies that γ is a $\boldsymbol{P}$−good curve and that $z(x) = Z_{\boldsymbol{P}}(\gamma)$, which yields the thesis.

11.4.2 Maximal Slope in the Network Direction

The next property that can be proven in general (i.e., under no extra assumption on α, X, μ ...) on the landscape function is the most important in view of its meaning in river basins applications. Our interest is in a continuous counterpart of the landscape function of [6]. What we actually need is a result concerning the fact that, on the points of the irrigation network $S_{\boldsymbol{P}}$, the direction of maximal slope of z is exactly the direction of the network. If a $\boldsymbol{P}$−good curve γ_0 is fixed, by the definition of z, for a.e. t_0 the derivative of z along the curve γ at the point $x_0 = \gamma_0(t_0)$ is exactly $|\gamma_0|_{t_0,\boldsymbol{P}}^{\alpha-1}$. This is the reason why we prove the following result. Notice that, as we said, in this continuous case the function z cannot be expected to be very regular, and in fact the maximal slope result we are going to prove involves differentiability in a very pointwise way but very weak as well.

Theorem 11.18. *Let $x_0 = \gamma_0(t_0)$, where γ_0 is a $\boldsymbol{P}$−good curve, t_0 a time with $t_0 \leq T(\gamma_0)$ and $\theta_0 := |\gamma_0|_{t_0,\boldsymbol{P}} > 0$. Then, for any $x \notin \gamma_0([0, t_0])$, we have*

$$z(x) \geq z(x_0) - \theta_0^{\alpha-1}|x - x_0| - o(|x - x_0|).$$

This corresponds to saying that the slope at x_0 in the direction of the network is actually the maximal slope at x_0.

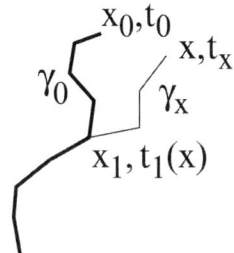

Fig. 11.2. Curves and points in the proof.

Proof. Let us fix $x \notin \gamma_0([0,t_0])$ such that $z(x) < z(x_0)$. We may assume that $x = \gamma_x(t_x)$ for a $\boldsymbol{P}$−good curve γ_x (otherwise $z(x) = +\infty$) and that the two curves γ_0 and γ_x get apart at a certain time $t_1(x) < t_0$ (the case $t_1(x) \geq t_0$ implies in fact $z(x) \geq z(x_0)$). By Lemma 11.19 (see below) we know that $t_1(x) \to t_0$ as $|x - x_0| \to 0$. Let us set $\theta(t) := |\gamma_0|_{t,\boldsymbol{P}}$ for $t \in [t_1(x), t_0]$ we may write $\theta(t) \leq \theta_0(1 + \varepsilon_x)$, where ε_x is infinitesimal as $|x - x_0| \to 0$ as a consequence of $t_1(x) \to t_0$. We use again Lemma 11.4 and its notations. In particular $A = [\gamma_0]_{t_0}$, $\theta_0 = \theta(t_0) = \boldsymbol{P}(A)$, $\mu' = \mu - \mu_A + \boldsymbol{P}(A)\delta_{x_0}$ and $\boldsymbol{P}' = \boldsymbol{P} - \mathbb{1}_A \cdot \boldsymbol{P} + \boldsymbol{P}(A)\delta_{\bar{\gamma}_0}$, where $\bar{\gamma}_0$ is the curve γ_0 stopped at time t_0 (see Figure 11.4.2). Let us also define, as in Theorem 11.5, $\mu'' = \mu - \mu_A + \boldsymbol{P}(A)\delta_x$ and $\boldsymbol{P}'' = \boldsymbol{P} - \mathbb{1}_A \cdot \boldsymbol{P} + \boldsymbol{P}(A)\delta_{\bar{\gamma}_x}$, where the curve $\bar{\gamma}_x$ is the curve γ_x stopped at time t_x. Notice that in $\boldsymbol{P}''$ we have eliminated the fibers that passed trough x_0 in the pattern $\boldsymbol{P}$ and we have redirected them to the curve $\bar{\gamma}_x$. The mass irrigated by this new pattern is $\pi_{\infty\sharp}\boldsymbol{P}'' = \mu''$. Then, by the optimality of $\boldsymbol{P}'$, we have

$$E^\alpha(\boldsymbol{P}') = X_\alpha(\mu') \leq X_\alpha(\mu'') + E^\alpha(\mu'', \mu') \leq E^\alpha(\boldsymbol{P}'') + |x - x_0|\theta_0^\alpha. \quad (11.9)$$

We want to compare $E^\alpha(\boldsymbol{P}')$ and $E^\alpha(\boldsymbol{P}'')$ and to do this here we need a more refined estimate than what we could find by using Theorem 11.2. As $\boldsymbol{P}' - \boldsymbol{P}'' = \theta_0(\delta_{\bar{\gamma}_0} - \delta_{\bar{\gamma}_x})$, we have in particular $|y|_{\boldsymbol{P}''} = |y|_{\boldsymbol{P}'} + \theta_0(\mathbb{1}_{y\in\bar{\gamma}_x} - \mathbb{1}_{y\in\bar{\gamma}_0})$. By using Formula (4.5) we get the following:

$$E^\alpha(\boldsymbol{P}'') - E^\alpha(\boldsymbol{P}') = \int_{\bar{\gamma}_x \setminus \bar{\gamma}_0} ((|y|_{\boldsymbol{P}'} + \theta_0)^\alpha - |y|_{\boldsymbol{P}'}^\alpha) \, d\mathcal{H}^1$$

$$- \int_{\bar{\gamma}_0 \setminus \bar{\gamma}_x} (|y|_{\boldsymbol{P}'}^\alpha - (|y|_{\boldsymbol{P}'} - \theta_0)^\alpha) \, d\mathcal{H}^1.$$

It is not difficult to check that, for $y \in \bar{\gamma}_x \cup \bar{\gamma}_0$, $|y|_{\boldsymbol{P}'} = |y|_{\boldsymbol{P}}$, as we have replaced the part of $\boldsymbol{P}$ concentrated on A by an equal amount of mass on $\bar{\gamma}_0$. Hence we may estimate (rewriting the integrals w.r.t. $\mathcal{H}^1$ as integrals in dt)

$$E^\alpha(\boldsymbol{P}'') - E^\alpha(\boldsymbol{P}') \leq \alpha \int_{t_1(x)}^{t_x} |\gamma_x(t)|_{\boldsymbol{P}}^{\alpha-1}\theta_0 \, dt - \int_{t_1(x)}^{t_0} (\theta(t)^\alpha - (\theta(t) - \theta_0)^\alpha) \, dt.$$

Since the function $s \mapsto s^\alpha - (s - \theta_0)^\alpha$ is decreasing and $\theta(t) \leq (1+\varepsilon_x)\theta_0$, we get $\theta(t)^\alpha - (\theta(t) - \theta_0)^\alpha \geq \theta_0^\alpha ((1 + \varepsilon_x)^\alpha - \varepsilon_x^\alpha)$. Hence we get

$$E^\alpha(\boldsymbol{P}'') - E^\alpha(\boldsymbol{P}') \leq \alpha\theta_0(z(x) - z(x_1)) - |t_0 - t_1(x)|\theta_0^\alpha ((1 + \varepsilon_x)^\alpha - \varepsilon_x^\alpha),$$

where $x_1 = \gamma_0(t_1(x)) = \gamma_x(t_1(x))$. Write $(1 + \varepsilon_x)^\alpha - \varepsilon_x^\alpha = (1 + \varepsilon_x')^{-1}$ and $\varepsilon_x' > 0$ is infinitesimal as $x \to x_0$. From $\theta_0^{\alpha-1} \geq (\theta(t))^{\alpha-1}$ we may estimate the difference $z(x_0) - z(x_1)$ by $z(x_0) - z(x_1) \leq |t_0 - t_1(x)|\theta_0^{\alpha-1}$. Now notice that, for $|x - x_0|$ sufficiently small, the inequality $\alpha < (1 + \varepsilon_x')^{-1}$ is satisfied, and hence

$$E^\alpha(\boldsymbol{P}'') \leq E^\alpha(\boldsymbol{P}') + (1 + \varepsilon_x')^{-1}\theta_0(z(x) - z(x_0)).$$

If we finally insert it into (11.9) we finally get

$$z(x) - z(x_0) \geq -\theta_0^{\alpha-1}|x - x_0|(1 + \varepsilon'_x).$$

Lemma 11.19. *According to the notations of Theorem 11.18 the parting time $t_1(x)$ tends to t_0 as $x \to x_0$ and $z(x) \leq z(x_0)$.*

Proof. Suppose, by contradiction, that there exists a sequence $x_k \to x_0$ such that $\lim_k t_1(x_k) = \bar{t} < t_0$ and $z(x_k) \leq z(x_0)$. Since γ_0 is injective (Corollary 11.7), we may infer the existence of a positive quantity δ such that $|\gamma_0(t_1(x_k)) - x_0| \geq \delta$ (otherwise there would be a time $t \leq \bar{t} < t_0$ with $\gamma_0(t) = x_0$). For any k consider a $\boldsymbol{P}$-good curve γ_k such that $x_k = \gamma_k(t_k)$. Let us consider the points $\gamma_k(\bar{t} + \delta/2)$: this collection of points must in fact be finite, otherwise we would have $|\gamma_k|_{\bar{t}+\delta/2,\boldsymbol{P}} \to 0$ and hence $z(x_k) \geq z(x_k) - z(\gamma_k(\bar{t} + \delta/2)) \geq |\gamma_k|_{\bar{t}+\delta/2,\boldsymbol{P}}^{\alpha-1}|t_k - (\bar{t} + \delta/2)| \to +\infty$ because $|t_k - (\bar{t} + \delta/2)| \geq |x_k - \gamma_0(\bar{t})| - 2\delta/3 \geq \delta/4$ for k large enough. This is in contradiction with $z(x_k) \leq z(x_0)$ and then we may suppose, up to subsequences, that $\gamma_k(\bar{t} + \delta/2) = \bar{x}$ (for a point $\bar{x}$ which does not belong to the image of γ_0, otherwise we would contradict Corollary 7.9) and that γ_k uniformly converges to a curve γ. At the limit we should get a curve γ passing through $\gamma_0(\bar{t})$, $\bar{x}$ and x_0, i.e. we have created a loop because γ_0 does not pass through $\bar{x}$. From $z(x_k) \leq z(x_0)$ we can infer by semicontinuity (Lemma 11.16) that γ is a $\boldsymbol{P}$-good curve and hence this loop is against Corollary 11.6 since $|\gamma_0|_{t_0,\boldsymbol{P}} > 0$.

11.5 Hölder Continuity under Extra Assumptions

Here we will be able to prove some extra regularity properties on z, but we have to add some assumptions. The most important ones are on α ($\alpha > 1 - 1/N$ is required) and on the irrigated measure μ (a lower bound on its density is supposed).

11.5.1 Campanato Spaces by Medians

We will here give a simple variant of a well-known result by Campanato (see [26]) about an integral characterization of Hölder continuous functions.

Definition 11.20. *Given a measurable function u on a domain U we call median of u in U any number m which satisfies the following equivalent conditions:*

- *$|\{x \in U : u(x) > m\}| \leq \frac{1}{2}|U|$ and $|\{x \in U : u(x) < m\}| \leq \frac{1}{2}|U|$;*
- *there exists a measurable subset $A \subset \{x \in U : u(x) = m\}$ such that $|\{x \in U : u(x) > m\} \cup A| = \frac{1}{2}|U|$;*
- *the function $t \mapsto \int_U |u(x) - t|dx$ achieves its minimum at $t = m$.*

The sets of medians of u in U is an interval of $\mathbb{R}$; the middle point of this interval is called central median of u in U

Definition 11.21. *If A is a given positive number, a domain $X \subset \mathbb{R}^N$ is said to be of type A if $|X_{x_0,r}| \geq Ar^N$ for any $r \in [0, \mathrm{diam}\, X]$ and for all $x_0 \in X$, where $X_{x_0,r} = X \cap B(x_0, r)$.*

Lemma 11.22. *If X is a domain of type A and u is a function in $L^1(X)$ such that for a finite constant C and any $r \in [0, \mathrm{diam}\, X]$,*

$$\int_{X_{x_0,r}} |u - \tilde{u}_{x_0,r}|dx \leq Cr^{N+\beta},$$

for all $x_0 \in X$, where $\tilde{u}_{x_0,r}$ is the central median of u on $X_{x_0,r}$, then u admits a representative which is Hölder continuous of exponent β.

Proof. This is nothing but the fact that Campanato spaces may be built by using medians instead of average values. See the proof of Theorem 1.2 at page 70 in [43] and adapt it. In fact it is easy to see that for each point x_0 the value $\tilde{u}_{x_0,r}$ converges as $r \to 0$ to a value $\tilde{u}(x_0)$ and that

$$|\tilde{u}(x) - \tilde{u}(y)| \leq C|x - y|^{\beta},$$

exactly as in the proof we mentioned. What we need to prove is that $\tilde{u}(x) = u(x)$ a.e.. This can be obtained in this way: let us denote the average value of u on $X_{x_0,r}$ by $\bar{u}_{x_0,r}$. Then

$$|\bar{u}_{x_0,r} - \tilde{u}_{x_0,r}| \leq |X_{x_0,r}|^{-1} \int_{X_{x_0,r}} |u(x) - \tilde{u}_{x_0,r}|dx$$

$$\leq |X_{x_0,r}|^{-1} \int_{X_{x_0,r}} |u(x) - \bar{u}_{x_0,r}|dx,$$

where the second inequality has been established as a consequence of the minimality property of the median. As at Lebesgue points the last expression tends to zero, this implies that the average $\bar{u}_{x_0,r}$ and the median $\tilde{u}_{x_0,r}$ share the same limit a.e. On the same points we also have $\bar{u}_{x_0,r} \to u(x_0)$, and this proves $\tilde{u}(x_0) = u(x_0)$ a.e..

11.5.2 Hölder Continuity of the Landscape Function

Theorem 11.23. *Suppose that X is a domain of type A for $A > 0$, that $\alpha > 1 - 1/N$ and that $\mu \in \mathcal{P}(X)$ is a probability measure such that the density of its absolutely continuous part is bounded from below by a positive constant. Then any landscape function z has a representative $\tilde{z}$ which is Hölder continuous of exponent $\beta = N(\alpha - (1 - 1/N))$.*

Proof. Let us fix a measure μ_1 and apply Theorem 11.14 to it and μ. By using the triangle inequality for E^α, we get

$$-E^\alpha(\mu, \mu_1) \le X_\alpha(\mu_1) - X_\alpha(\mu) \le \alpha \int_X z \, d(\mu_1 - \mu), \qquad (11.10)$$

provided μ_1 is a measure of the form allowed in Theorem 11.14, i.e. $\mu_1 \ll \mu$ with bounded density. From (11.10) we get

$$\alpha \int_X z \, d(\mu - \mu_1) \le E^\alpha(\mu, \mu_1). \qquad (11.11)$$

Suppose that μ has an absolutely continuous part with density everywhere larger than $\lambda_0 > 0$ and choose

$$\mu_1 = \mu - \lambda_0 \mathbb{1}_A \cdot \mathcal{L}^N + \lambda_0 \mathbb{1}_B \cdot \mathcal{L}^N,$$

where A and B are two measurable subsets of $X_{x_0,\varepsilon}$ with $x_0 \in X$, $|A| = |B|$, $A \cup B = X_{x_0,\varepsilon}$ and $A \subset \{z \ge m\}$ and $B \subset \{z \le m\}$ and m is the central median value for z in $X_{x_0,\varepsilon}$. By construction μ_1 is a probability measure to which the estimate of Theorem 11.14 may be applied. With this choice of μ and μ_1 we get

$$\int_X z \, d(\mu - \mu_1) = \int_A z(x) \, \lambda_0 \, dx - \int_B z(x) \, \lambda_0 \, dx$$
$$= \lambda_0 \left(\int_A (z(x) - m) dx - \int_B (z(x) - m) dx \right) = \lambda_0 \int_{X_{x_0,\varepsilon}} |z(x) - m| dx.$$

Putting into (11.11)

$$\int_{X_{x_0,\varepsilon}} |z(x) - m| dx \le (\alpha \lambda_0)^{-1} E^\alpha(\mu, \mu_1).$$

To estimate $E^\alpha(\mu, \mu_1)$ use (6.3) and get

$$\int_{X_{x_0,\varepsilon}} |z(x) - m| dx \le \frac{C_{\alpha,N}}{\lambda_0^{1-\alpha}} \varepsilon^{1+\alpha N}.$$

Since $1 + \alpha N = N + \beta$, Lemma 11.22 may be applied.

An important consequence of this fact is the following:

Corollary 11.24. *Under the same assumptions on X, α and μ of Theorem 11.23, the inequality*

$$X_\alpha(\mu_1) \le X_\alpha(\mu) + \int_X \tilde{z} d(\mu_1 - \mu)$$

holds for any measure $\mu_1 \in \mathcal{P}(X)$.

Proof. The inequality holds for μ_1 of the form $\mu_1 = (1+g) \cdot \mu$ with $g \in L^\infty$, but any measure $\mu_1 \in \mathcal{P}(X)$ may be approximated by these kind of measures. Since $\tilde{z}$ is continuous, at both the sides of the inequalities we have quantities which are continuous with respect to weak convergence in the variable μ_1. This allows to conclude that the same inequality is valid for any μ_1.

Even if we have proven that the landscape function z equals a.e. a function which is Hölder continuous, this is not enough. In fact, this result does not provide information on the behavior of z on negligible sets. Yet, the pointwise values of z on $S_{\mathbf{P}}$ are of particular interest (as in last Section), and $S_{\mathbf{P}}$ is one-dimensional and thus negligible. This is why the next step will be proving that z and $\tilde{z}$ actually agree everywhere.

Theorem 11.25. *Let m_ε denote the central median of z in the ball $B(x_0, \varepsilon)$. Under the same assumptions of Theorem 11.23 one has $m_\varepsilon \to z(x_0)$ as $\varepsilon \to 0$. Consequently, we have $\tilde{z}(x_0) = z(x_0)$.*

Proof. By the semicontinuity of z it is easy to get $\liminf_{\varepsilon \to 0} m_\varepsilon \geq z(x_0)$, hence only an estimate from above for m_ε is needed. Let us now consider a ball $B(x_0, \varepsilon)$ and a set $A_\varepsilon \subset B(x_0, \varepsilon) \cap \{z \geq m_\varepsilon\}$ such that $|A_\varepsilon| = |X_{x_0, \varepsilon}|/2$. Then set $K_\varepsilon = \{\gamma \in K : (\pi_\infty)(\gamma) \in A_\varepsilon\}$, $\mu_\varepsilon = \mu + \mu(A_\varepsilon)\delta_{x_0} - \mathbb{1}_{A_\varepsilon} \cdot \mu$, and $\mathbf{P}_\varepsilon = \mathbf{P} + \mathbf{P}(K_\varepsilon)\delta_{\gamma_0} - \mathbb{1}_{K_\varepsilon} \cdot \mathbf{P}$, where γ_0 is an $\mathbf{P}$-good curve stopping at x_0. Theorem 11.2 can be applied to $\mathbf{P}$ and $\mathbf{P}_\varepsilon$ and hence we have

$$X_\alpha(\mu_\varepsilon) \leq E^\alpha(\mathbf{P}_\varepsilon) \leq E^\alpha(\mathbf{P}) + \alpha \left(\mathbf{P}(K_\varepsilon) Z_{\mathbf{P}}(\gamma_0) - \int_{K_\varepsilon} Z_{\mathbf{P}} d\mathbf{P} \right)$$

$$= E^\alpha(\mathbf{P}) + \alpha\mu(A_\varepsilon) z(x_0) - \alpha \int_{A_\varepsilon} z(x)\mu(dx)$$

$$\leq E^\alpha(\mathbf{P}) + \alpha\mu(A_\varepsilon)(z(x_0) - m_\varepsilon)$$

$$\leq X_\alpha(\mu) + \alpha\mu(A_\varepsilon)(z(x_0) - m_\varepsilon).$$

Hence, by Corollary 6.7, we have

$$X_\alpha(\mu) \leq X_\alpha(\mu_\varepsilon) + C\varepsilon\mu(A_\varepsilon)^\alpha \leq X_\alpha(\mu) + \alpha\mu(A_\varepsilon)(z(x_0) - m_\varepsilon) + C\varepsilon\mu(A_\varepsilon)^\alpha.$$

This implies

$$m_\varepsilon - z(x_0) \leq C\varepsilon\mu(A_\varepsilon)^{\alpha-1} \leq C\varepsilon^{1+N(\alpha-1)}.$$

Since the exponent $1 + N(\alpha-1)$ is larger than 0 we get $\limsup_{\varepsilon \to 0} m_\varepsilon \leq z(x_0)$. To get the second part of the thesis, just use $\tilde{z}(x_0) = \lim_{\varepsilon \to 0} m_\varepsilon$.

Remark 11.26. The landscape function z is in general never Lipschitz continuous (not even locally), as on the set $S_{\mathbf{P}}$ it has slopes given by $\theta^{\alpha-1}$. This means that, if we have arbitrarily small values of θ, we cannot have a Lipschitz constant for z. Yet estimates of the kind $\theta \geq c > 0$ would imply $\mathcal{H}^1(S_{\mathbf{P}}) < +\infty$ and no measure whose support is not one-dimensional may be irrigated by a set of finite length (or locally finite length).

12 The Gilbert-Steiner Problem

In this chapter, we consider the irrigation problem for two finite atomic measures μ^+ and μ^-. This problem was first set and studied in a finite graph setting by Gilbert in [44]. Following his steps, we first consider the irrigation problem from a source to two Dirac masses. If the optimal structure is made of three edges, the first order condition for a local optimum yields constraints on the angles between the edges at the bifurcation point (see Lemma 12.2). Conversely, thanks to the central angle property, Lemma 12.6 describes how these angle constraints permit to construct the bifurcation point directly from the source and the two masses. Proposition 12.10 then completely describes and proves the structure of an optimum for given locations and weights of the source and the two Dirac masses. Let us mention that the classification of the different possible optimal structures is given in Gilbert's article, yet without proof. In the second section, we generalize the construction already made for two masses to any number of masses. More precisely, if we prescribe a particular topology of a tree irrigating n masses from a Dirac mass, we describe a recursive procedure that permits to construct a local optimum that has this topology. Finally, in the last section, we raise the question of the number of edges meeting at a bifurcation point in an optimal traffic plan with graph structure. In particular, Proposition 12.17 proves that for $\alpha \leq \frac{1}{2}$ and for an optimal traffic plan in $\mathbb{R}^2$, there cannot be more than three edges meeting at a bifurcation point. It is an open question and a numerically plausible conjecture that the result holds for general α. The presentation follows closely Bernot's PhD [10]. The main results come from this thesis and from the original Gibert paper [44].

12.1 Optimum Irrigation from One Source to Two Sinks

Let A_1, A_2, A_3 be three distinct points in $\mathbb{R}^N$, $\mu^- = m_1\delta_{A_1} + m_2\delta_{A_2}$ and $\mu^+ = m_3\delta_{A_3}$ with $m_3 = m_1 + m_2$ and $m_1, m_2 > 0$.

Lemma 12.1. *In the case A_1, A_2, A_3 are aligned, an optimal traffic plan from μ^+ to μ^- has its support in the minimal segment containing A_1, A_2, A_3. If A_1, A_2, A_3 are not aligned, an optimal traffic plan has its support in the triangle A_1, A_2, A_3. In addition, it is a graph with two or three edges.*

M. Bernot et al., *Optimal Transportation Networks*. Lecture Notes
in Mathematics 1955.
© Springer-Verlag Berlin Heidelberg 2009

Proof. Because of the convex envelop property 5.15, the support of an optimal traffic plan from μ^+ to μ^- is in the convex envelop of A_1, A_2 and A_3. Further, Proposition 9.1 proves that an optimal traffic plan is a graph with at most three edges.

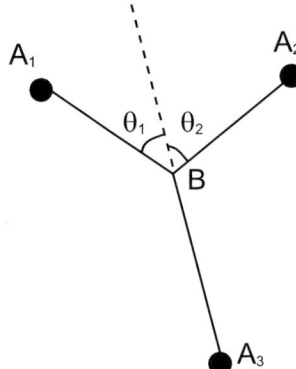

Fig. 12.1. If an optimum has a Y-structure, the perturbation of the bifurcation point gives necessary condition through the cancellation of the derivative of the cost. This prescribes the angles θ_1 and θ_2.

Lemma 12.2. *Let P be an optimal traffic plan from μ^+ to μ^- made of three edges. With the notation of Figure 12.1, the bifurcation point B has to satisfy the following angle constraints:*

$$\cos(\theta_1) = \frac{k_1^{2\alpha} + 1 - k_2^{2\alpha}}{2k_1^{\alpha}} \tag{12.1}$$

$$\cos(\theta_2) = \frac{k_2^{2\alpha} + 1 - k_1^{2\alpha}}{2k_2^{\alpha}} \tag{12.2}$$

$$\cos(\theta_1 + \theta_2) = \frac{1 - k_2^{2\alpha} - k_1^{2\alpha}}{2k_1^{\alpha}k_2^{\alpha}}, \tag{12.3}$$

where $k_1 = \frac{m_1}{m_1 + m_2}$, $k_2 = \frac{m_2}{m_1 + m_2}$.

Proof. Because of Lemma 12.1, it is equivalent to consider the two dimension situation. Let us consider the graph $G(B)$ made of oriented weighed edges (A_1B, m_1), (A_2B, m_2) and (BA_3, m_3), with $B \in \mathbb{R}^2 \setminus \{A_1, A_2, A_3\}$. The cost of this graph is

$$E^{\alpha}(G(B)) = m_1^{\alpha}|A_1B| + m_2^{\alpha}|A_2B| + m_3^{\alpha}|BA_3|.$$

Notice that this function of B is differentiable on $\mathbb{R}^2 \setminus \{A_1, A_2, A_3\}$. Thus, if $G(B)$ is an optimal traffic plan with $B \notin \{A_1, A_2, A_3\}$, the first derivative of $E^{\alpha}(G(B))$ has to cancel.

For $B \notin \{A_1, A_2, A_3\}$, let us denote by $\mathbf{n}_i = \frac{A_iB}{||A_iB||}$ the unit vector from A_i to B for $i = 1, 2, 3$. Since

$$||A_iB + \mathbf{v}|| = ||A_iB|| + \mathbf{v} \cdot \mathbf{n}_i + o(||\mathbf{v}||),$$

the necessary condition given by the cancellation of the derivative of the cost function yields the balance equation

$$m_1^\alpha \mathbf{n}_1 + m_2^\alpha \mathbf{n}_2 + m_3^\alpha \mathbf{n}_3 = 0. \tag{12.4}$$

Let θ_i be the angle between $\mathbf{n}_i$ and $-\mathbf{n}_3$ for $i = 1, 2$ and $k_1 = \frac{m_1}{m_1+m_2}$, $k_2 = \frac{m_2}{m_1+m_2}$. Multiplying the balance equation (12.4) by $\mathbf{n}_i$ for $i = 1, 2, 3$ we obtain the following equalities:

$$k_1^\alpha + k_2^\alpha \mathbf{n}_1 \cdot \mathbf{n}_2 = \cos(\theta_1) \tag{12.5}$$
$$k_1^\alpha \mathbf{n}_1 \cdot \mathbf{n}_2 + k_2^\alpha = \cos(\theta_2) \tag{12.6}$$
$$k_1^\alpha \cos(\theta_1) + k_2^\alpha \cos(\theta_2) = 1, \tag{12.7}$$

so that the angles satisfy

$$\cos(\theta_1) = \frac{k_1^{2\alpha} + 1 - k_2^{2\alpha}}{2k_1^\alpha} \tag{12.8}$$

$$\cos(\theta_2) = \frac{k_2^{2\alpha} + 1 - k_1^{2\alpha}}{2k_2^\alpha} \tag{12.9}$$

$$\cos(\theta_1 + \theta_2) = \frac{1 - k_2^{2\alpha} - k_1^{2\alpha}}{2k_1^\alpha k_2^\alpha}. \tag{12.10}$$

This means that in the triangle A_1BA_3, the angle at B is $\pi - \theta_1$ and in the triangle A_2BA_3, the angle at B is $\pi - \theta_2$.

Remark 12.3. Notice that in the case $m_1 = m_2$,

$$\theta_1 = \theta_2 = \arccos(2^{2\alpha-1} - 1)/2.$$

If $\alpha = \frac{1}{2}$ the angles satisfy $\theta_1 + \theta_2 = \frac{\pi}{2}$, $\theta_1 = \sqrt{k_1}$ and $\theta_2 = \sqrt{k_2}$. Thus the bifurcation point lies on the circle of diameter A_1A_2. If $\alpha = 0$, we find the $\frac{2\pi}{3}$ angle constraint that has to satisfy a Steiner point in the Steiner tree problem.

As we shall see in the following, the angle constraints permit to locate precisely the branching point. In doing so, the central angle property plays a major role.

Lemma 12.4. *(Central angle property, see Figure 12.2) Let A_1 and A_2 be two points on a circle centered at O, and B on the largest of the two circle arcs A_1A_2. Then the angle A_1OA_2 is twice the angle A_1BA_2.*

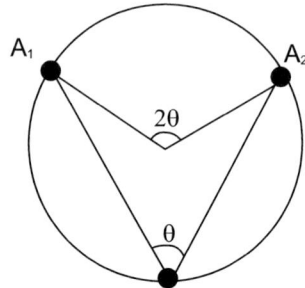

Fig. 12.2. The central angle property states that the central angle is twice the angle $A_1 B A_2$.

Lemma 12.5. *Given two points A_1 and A_2 and an angle $\theta \leq \frac{\pi}{2}$, the set of points B such that the not oriented angle $A_1 B A_2$ is θ is the union of two circle arcs going through A_1 and A_2, with radius $\frac{|A_1 A_2|}{2\sin(\theta)}$. We shall call these circle arcs, "equiangle arcs".*

Proof. Let O_1 and O_2 be the two points equidistant from A_1 and A_2 and such that they intercept the arc $A_1 A_2$ with an angle 2θ. The central angle property (see Lemma 12.4) applied to the circle centered at O_1 with radius $O_1 A_1$ states that the largest circle arc $A_1 A_2$ is made of points B such that the angle $A_1 B A_2$ is θ. The same property holds for the largest circle arc $A_1 A_2$ of the circle centered at O_2. Conversely every point B lies on such a circle arc. Indeed, let B be such that the angle $A_1 B A_2$ is θ. Since $\theta \leq \frac{\pi}{2}$, the circle going through B, A_1 and A_2 is such that B is on the largest circle arc $A_1 A_2$. Thus, its center O is such that the angle $A_1 O A_2$ is 2θ so that O is either O_1 or O_2. Finally, the radius of the two possible circle arcs is the distance $O_1 A_1$, i.e. $\frac{|A_1 A_2|}{2\sin(\theta)}$.

Lemma 12.6. *Let $\boldsymbol{P}$ be an optimal traffic plan from μ^+ to μ^- made of three edges. Let $\mathcal{C}$ be the equiangle arc associated to A_1, A_2 and $\theta := \theta_1 + \theta_2$, which is in the same half plane as A_3. Let $\mathcal{C}'$ be the complementary circle arc of $\mathcal{C}$. There is a "pivot" point $A_{1,2} \in \mathcal{C}'$ which does not depend on A_3 and such that the bifurcation point B is the intersection of $A_3 A_{1,2}$ with $\mathcal{C}$.*

Proof. Let us denote by $A_{1,2}$ the intersection of the line $A_3 B$ with $\mathcal{C}'$. Since $\boldsymbol{P}$ is supposed to be optimal, the bifurcation point B has to satisfy the angle conditions given by Lemma 12.2, i.e. the angle $A_{1,2} B A_1$ is prescribed as equal to θ_1. Thus, the largest arc $A_1 A_{1,2}$ is an equiangle arc for the angle θ_1. Let us denote by O the center of the equiangle arc. Because of the central angle property (see Lemma 12.4), the angle $A_1 O A_{1,2}$ is twice the angle $A_1 B A_{1,2}$. Thus, the point $A_{1,2}$ is easily constructed as the image of A_1 under the rotation of angle $2\theta_1$ centered at O. In particular, the point $A_{1,2}$ does not depend on the source point A_3 and the bifurcation point B (see Figure 12.3) is obtained as the intersection of the line $A_3 A_{1,2}$ with $\mathcal{C}$.

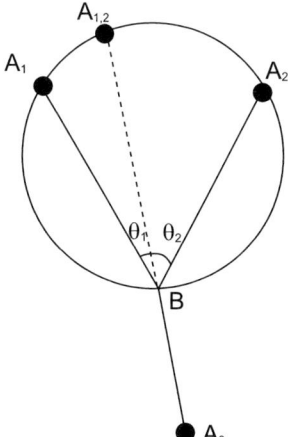

Fig. 12.3. The equiangle locus associated to $A_1, A_{1,2}$ with an angle θ_1 is supported by a circle and thus is the same circle as the equiangle locus associated to A_1, A_2 with an angle $\theta_1 + \theta_2$. The bifurcation point can then be obtained as the intersection of $A_1 A_3$ with the circle arc $A_1 A_2$.

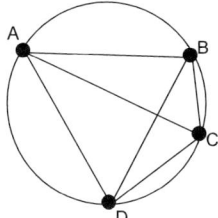

Fig. 12.4. Ptolemy's theorem asserts that if A, B, C, D are cocyclic, the product of the two diagonals is equal to the sum of the products of the opposite sides.

To conclude the study of optimal structures from on source to two sinks, we need to use the following classical geometric inequality.

Theorem 12.7. *(Ptolemy's theorem, see Figure 12.4) Let A, B, C, D be four points such that $ABCD$ is a convex quadrilateral. Then*

$$|AC||BD| + |AB||CD| \geq |AD||BC|,$$

and the equality holds if and only if A, B, C, D are cocylic.

Lemma 12.8. *Let P be an optimal traffic plan from μ^+ to μ^- made of three edges and $A_{1,2}$ the "pivot" point obtained in Lemma 12.6. Then, $E^\alpha(P) = m_3^\alpha |A_3 A_{1,2}|.$*

Proof. Indeed, it is a direct consequence of Ptolemy's theorem (see Theorem 12.7) stating that the product of the diagonals of a quadrilateral whose vertices lie in a circle equals the sum of the products of the opposite sides. Let us notice first that $|A_1 A_{1,2}| = |A_1 A_2|(\frac{m_2}{m_3})^\alpha$ and $|A_2 A_{1,2}| = |A_1 A_2|(\frac{m_1}{m_3})^\alpha$. When applied to the quadrilateral $A_1 B A_2 A_{1,2}$, the theorem becomes $|B A_{1,2}||A_1 A_2| = |B A_1||A_{1,2} A_2| + |B A_2||A_{1,2} A_1|$. Thus,

$$
\begin{aligned}
E^\alpha(P) &= m_1^\alpha |B A_1| + m_2^\alpha |B A_2| + m_3^\alpha |A_3 B| \\
&= m_3^\alpha (|B A_{1,2}| + |A_3 B|) \\
&= m_3^\alpha |A_3 A_{1,2}|.
\end{aligned}
$$

Definition 12.9. *Let P be an optimal traffic plan from μ^+ to μ^-. Lemma 12.1 states that P is either made of two or three edges. We consider three possible optimal structures:*

- *The "Y" structure in the case P has three edges. Notice that the location of the bifurcation point is given by Lemma 12.6.*
- *The "V" structure, i.e. P is made of the edges $(A_3 A_1, m_1)$ and $(A_3 A_2, m_2)$.*
- *Two possible "L" structures, i.e. P is made of the edges $(A_3 A_1, m_3)$ and $(A_1 A_2, m_2)$ or $(A_3 A_2, m_3)$ and $(A_2 A_1, m_1)$.*

Proposition 12.10. *Let P be an optimal traffic plan from μ^+ to μ^-. Let $A_{1,2}$ be the pivot point associated to $(A_1, m_1), (A_2, m_2)$, in the half plane not containing A_3. Let C be the equiangle circle arc that in the same half plane as A_3. There are four different zones for A_3.*

- *If $A_3 A_{1,2} \cap C = \{B\}$ and $B \in [A_3 A_{1,2}]$, then P has a "Y" structure.*
- *If $A_3 A_{1,2} \cap C = \{B\}$ and $B \notin [A_3 A_{1,2}]$, then P has a "V" structure.*
- *If $A_3 A_{1,2} \cap C = \emptyset$, then P has an "L" structure. More precisely, either $|A_3 A_1| < |A_3 A_2|$ and P is made of the two edges $(A_3 A_1, m_3)$ and $(A_1 A_2, m_2)$ or $|A_3 A_2| < |A_3 A_1|$ and it is made of the two edges $(A_3 A_2, m_3)$ and $(A_2 A_1, m_1)$.*

Proof. Because of the central angle property, the angles $A_2 A_1 A_{1,2}$ and $A_1 A_2 A_{1,2}$ equal to θ_2 and θ_1, respectively. After the suitable rescaling, we can suppose that the distance $|A_1 A_2| = 1$, and $m_3 := m_1 + m_2 = 1$. In that case, the distances $|A_1 A_{1,2}|$ and $|A_2 A_{1,2}|$ are respectively equal to m_2^α and m_1^α. For the sake of simplicity, we shall denote $a = |A_3 A_1|$, $b = |A_3 A_2|$, $c = |A_1 A_{1,2}|$ and $d = |A_2 A_{1,2}|$. The four different zones for A_3 are illustrated by Figure 12.7.

If $B \in (A_3 A_{1,2}]$, Lemma 12.8 states that the "Y" structure with a bifurcation at B has a cost $|A_3 A_{1,2}|$. The three possible competitors are the "V" structure and the two possible "L" structures with costs respectively equal to $ad + bc$, $a + c$ and $b + d$. Because of the triangular inequality, $a + c > |A_3 A_{1,2}|$ and $b + d > |A_3 A_{1,2}|$ so that an "L" structure is not optimal (see Figure 12.5.) Since A_3 is not cocyclic with $A_1, A_2, A_{1,2}$, Ptolemy's theorem ensures that

$$
ad + bc > |A_3 A_{1,2}||A_1 A_2|,
$$

so that the "V" structure is not optimal either.

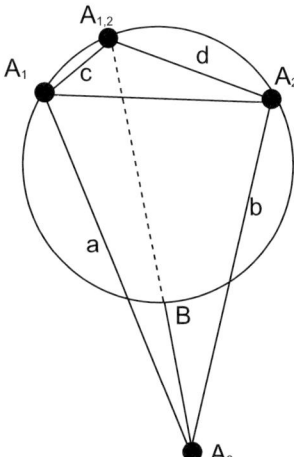

Fig. 12.5. See the proof of Proposition 12.10. If $B \in (A_3 A_{1,2}]$, the cost of the "Y" structure is $|A_3 A_{1,2}|$. By triangular inequality, it is greater than $a + c$ and $b + d$. By Ptolemy's theorem, it is also greater than $ad + bc$. Thus, the "Y" structure is optima.

If $B \notin [A_3 A_{1,2}]$ or $A_3 A_{1,2} \cap \mathcal{C} = \emptyset$, then an optimal structure cannot have three edges because no bifurcation point is able to satisfy necessary angle conditions. An optimum must have an "L" or "V" structure, depending on the position of the source point A_3.

Let us first consider the case of Figure 12.6, where $A_3 A_{1,2} \cap \mathcal{C} = \emptyset$. This means that the source point A_3 is either located in the zone delimited by the lines $A_2 A_1$ and $A_{1,2} A_1$, or in the zone delimited by $A_1 A_2$ and $A_{1,2} A_2$. Without loss of generality we assume that we are in the first case. This can be translated by saying that the angle θ between the line $A_1 A_2$ and $A_3 A_1$ is lower than θ_2 (in the second case the angle θ between the lines $A_1 A_2$ and $A_3 A_2$ is lower than θ_1). We want to prove that the "L" structure is better than the "V" structure, i.e.

$$a + c < ad + bc = ad + c\sqrt{a^2 + 1 + 2a \cos(\theta)}.$$

Since $\cos(\theta) \geq \cos(\theta_2)$, it is sufficient to prove that

$$ad + c\sqrt{a^2 + 1 + 2a \cos(\theta_2)} > a + c,$$

which is equivalent to

$$\cos(\theta_2) > \frac{1 - d}{c} - \frac{a}{2}\left(1 - \left(\frac{1 - d}{c}\right)^2\right).$$

Since $\frac{1-d}{c} < 1$ and $\cos(\theta_2) = \frac{1+c^2-d^2}{2c} > \frac{1-d}{c}$, we obtain the claimed inequality.

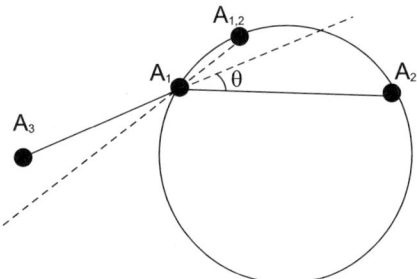

Fig. 12.6. If $A_3 A_{1,2} \cap \mathcal{C} = \emptyset$, the angle θ is lower than θ_2. This permits to prove that the "L" structure is better than the "T" structure.

Let us finally consider the case when $B \notin [A_3 A_{1,2}]$, i.e. the case when the source point A_3 is such that the angle $A_1 A_3 A_2$ is greater than $\theta_1 + \theta_2$. We want to prove that the "V" structure is better than both "L" structures. That is to say, we have to prove that $ad + bc < a + c$ and $ad + bc < b + d$. Permuting the masses of A_1 and A_2 amounts to exchange the role of the two points, so that it is sufficient to prove that

$$ad + bc < a + c.$$

We notice first that this inequality is satisfied for any point on $\mathcal{C} \cup [A_1, A_2] \setminus A_1$. Indeed, because of Ptolemy's theorem and the triangular inequality, if $A_3 \in \mathcal{C} \setminus A_1$, $ad + bc = |A_3 A_{1,2}| < a + c$. If $A_3 \in (A_1, A_2]$, then $0 \le b < 1$ and $d < 1$, so that $ad + bc < a + c$. Let us suppose that for some point $\tilde{A}_3$, at distance $\tilde{a}$ and $\tilde{b}$ of A_1 and A_2, the equality $\tilde{a}d + \tilde{b}c = \tilde{a} + c$ is satisfied. Then, since $ad + bc - a - c$ is affine in (a, b), the whole half-line

$$(a, b) = (\tilde{a} + \lambda c, \tilde{b} + \lambda(1 - d)),$$

where $\lambda \ge 0$ satisfies the equality. Let us see under what conditions on λ there is indeed a source point A_3 such that $|A_3 A_1| = a$ and $|A_3 A_2| = b$. This amounts to prove that $1 - b < a < 1 + b$, i.e. the circles with radius a and b and centered respectively at A_1 and A_2 intersect. Observe that we have

$$a + b = \tilde{a} + \tilde{b} + \lambda(c + 1 - d) > \tilde{a} + \tilde{b} > 1$$

for all $\lambda \ge 0$. In addition, we have

$$a - b = \tilde{a} - \tilde{b} + \lambda(c + d - 1),$$

where $\tilde{a} - \tilde{b} < 1$ and $c + d - 1 > 0$. Thus, there is some value $\lambda_m > 0$ such that

$$\tilde{a} - \tilde{b} + \lambda_m(c + d - 1) = 1.$$

For $0 \leq \lambda < \lambda_m$, we have $a < 1 + b$, so that there is a corresponding source point A_3 such that $|A_3 A_1| = a$ and $|A_3 A_2| = b$. For the limit case $\lambda = \lambda_m$, we have $a = b + 1$ so that the corresponding source point A_3 is colinear with A_1 and A_2 and outside the segment $[A_1, A_2]$. Thus, if we consider the curve of points A_3 that is obtained when λ goes from 0 to λ_m, this curve connects the point $\tilde{A}_3$ to a point on the line $A_1 A_2$. Thus, by continuity, there is some $\lambda_0 \in (0, \lambda_m)$ such that the source point A_3 corresponding to $(\tilde{a} + \lambda_0 c, \tilde{b} + \lambda_0(1 - d))$ lies on $\mathcal{C} \cup [A_1, A_2]$. This particular point A_3 cannot be A_1 because $\tilde{a} + \lambda_0 c > 0$. Thus $A_3 \in \mathcal{C} \cup [A_1, A_2] \setminus A_1$ which is a contradiction because we saw that $ad + bc < a + c$ on this set. This means that no point $\tilde{A}_3$ is such that $ad + bc = c + a$ within the portion of disk delimited by the equiangle arc so that $ad + bc < a + c$ for all A_3 in this zone.

12.2 Optimal Shape of a Traffic Plan with given Dyadic Topology

The irrigation problem can be divided into two optimization problems: the optimization of the topology, and the optimization of the locations of bifurcation points. The optimization of the topology is treated in Chapter 13 in the particular case of the irrigation of the Lebesgue segment from a source. Within this section, we present a recursive construction that gives the location of bifurcation points (i.e. Steiner points) of an optimal structure that has a prescribed topology in two dimensions. To explain this construction, we consider the simpler case of trees with full Steiner topology.

12.2.1 Topology of a Graph

Definition 12.11. *A topology $\mathcal{T}$ for a given point set $(v_i)_{i=1}^n$ of $\mathbb{R}^N$ is an undirected connected graph $G = (V, E)$ where E is the set of edges and $V = (v_i)_{i=1}^{n+m}$ is the set of vertices. The points $(v_i)_{i=n+1}^{n+m}$ which are not present in the initial point set $(v_i)_{i=1}^n$ are called Steiner points.*

Definition 12.12. *A finite traffic plan induces a graph structure and thus a topology. Let us denote $\mathrm{TP}(\mu^+, \mu^-, \mathcal{T})$ the set of traffic plans with topology $\mathcal{T}$ and*

$$C(\mu^+, \mu^-, \mathcal{T}) := \inf_{\mathrm{TP}(\mu^+, \mu^-, \mathcal{T})} E^\alpha(\boldsymbol{P})$$

the cost of the topology $\mathcal{T}$.

Definition 12.13. *A Steiner topology is a topology $\mathcal{T}$ such that all vertices corresponding to Steiner point have degree 3. A full Steiner topology is such that it has $2n - 2$ vertices $(v_i)_{i=1}^{2n-2}$ and $2n - 3$ edges.*

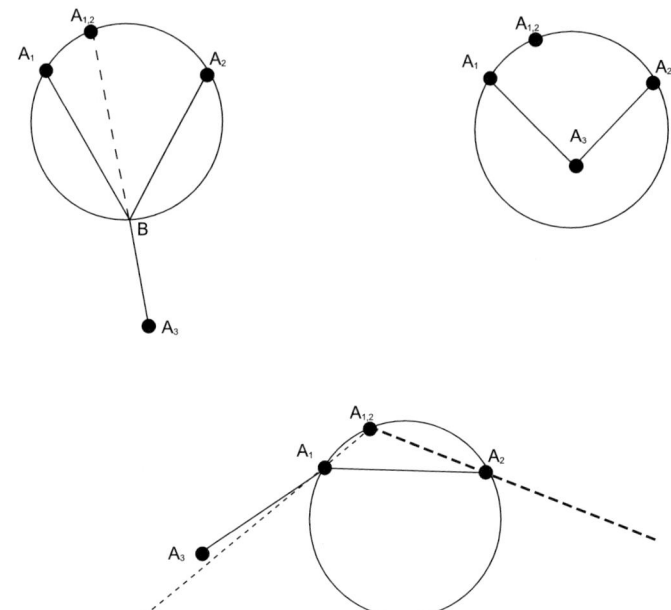

Fig. 12.7. Let us sum up the process that permits to find the optimal structure from one source to two sinks. 1) Given the masses m_1 and m_2, we obtain the angle θ at an optimal bifurcation. 2) We draw the equiangle arc, the pivot point and the line $A_3A_{1,2}$. 3) Depending on the position of the intersection between this line and the equiangle circle arc, we obtain one of the four possible configuration. Three of them are represented on this figure.

12.2.2 A Recursive Construction of an Optimum with Full Steiner Topology

In this subsection, we shall assume that the optimal structure associated to a prescribed full Steiner topology is not degenerated.

Let us first recall the construction of the optimal structure in the case we transport a source S to two Dirac masses located at points A and B. Proposition 12.10 states that there is a pivot point P such that the only bifurcation point of the optimal structure with full Steiner topology is obtained as the intersection of the line SP with the circle going through A, B and P.

In the more general case, this construction can be applied recursively as it was first described by Gilbert in [44]. Let us explain the recursive construction on a simple example. We consider μ^- as a target measure made of 4 Dirac masses A, B, C, D, and μ^+ the Dirac mass at a source point S. Let us suppose that the optimal structure has the full Steiner topology such that the first bifurcation occurs at B_1 and the first subtree irrigates A and B and the second subtree irrigates C and D. The second bifurcation at B_2 is such that one branch irrigates A and the other one B. At last, the bifurcation B_3 is such that one branch irrigates C and the other one D. This topology is in fact the simplest we can imagine and is illustrated by Figure 12.9.

Let us explain why the construction of bifurcation points B_i is only a recursive way to apply the construction in the simplest "one source to 2 Dirac masses" case. Indeed, if we look for the best structure, every subtree has to be optimal for the irrigation problem it induces. That is to say, the subtree which irrigates A and B from B_1 is optimal, so is the subtree irrigating C and D from B_1. Thus, points B_2 and B_3 can be constructed thanks to pivot points P_1 and P_2 as in Proposition 12.10. Next, the irrigation from S to B_2 and B_3 has to be optimal as a subtree of an optimal structure. As a consequence, the irrigation from S to P_1 and P_2 is also optimal. Indeed, since B_1, B_2 and P_1 are aligned, and B_1, B_3, P_2 are also aligned, the angle $P_1 B_1 P_2$ is the angle prescribed by necessary conditions of Lemma 12.2. Thus, the transport from S to P_1 and P_2 is optimal and we can construct the position of B_1 through the pivot point P_3 associated to P_1 and P_2.

Let us now give the construction top to down then bottom-up.

- The prescribed topology is such that A is grouped with B and C with D. Thus we construct their associated pivot points P_1 and P_2.
- Since (to be found) bifurcation points B_2 and B_3 are then grouped, we construct the pivot point associated to P_1 and P_2.
- Since the subtree made of edges SB_3, $B_3 P_1$ and $B_3 P_2$ is optimal, the bifurcation point B_3 is obtained as the intersection of the line SP_3 with the circle $P_3 P_1 P_2$.
- Now that the bifurcation point B_3 is located, we obtain the bifurcation point B_1 as the intersection of the line $B_3 P_1$ with the circle $P_1 AB$. And we obtain the bifurcation point B_2 as the intersection of the line $B_3 P_2$ with the circle $P_2 CD$.

12.3 Number of Branches at a Bifurcation

In this section we investigate the geometry of branches at a bifurcation point of an optimal traffic plan. The optimal structure of a traffic plan from one Dirac mass to two Dirac masses is essential in all that follows. Lemmas 12.15 and 12.16 give lower bound (depending on α) on the angle between two edges starting from the same point. As a consequence, we prove that it is not possible for an optimal finite traffic plan in $\mathbb{R}^2$ to have more than three edges meeting at a bifurcation point (away from μ^+ and μ^-), when $\alpha \leq \frac{1}{2}$ (Proposition 12.17). It is still a conjecture whether or not four edges can meet at a bifurcation point in $\mathbb{R}^2$ when $1 > \alpha > \frac{1}{2}$, though numerical experiments seem to exclude this situation.

Lemma 12.14. *The function g defined by $g(m) = \frac{(m+1)^{2\alpha} - m^{2\alpha} - 1}{2m^\alpha}$ is nondecreasing on $]0,1]$ for $\frac{1}{2} < \alpha \leq 1$ and nonincreasing for $\alpha < \frac{1}{2}$. Thus,*

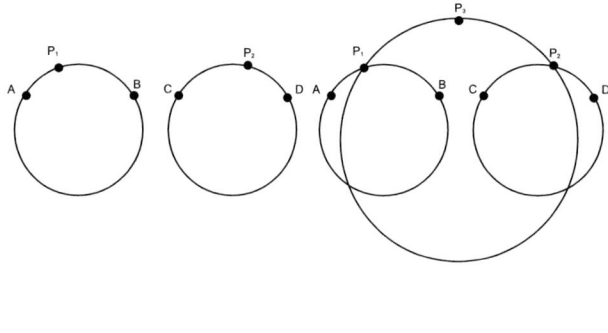

Fig. 12.8. Given a topology, the pivot point permits to reduce two masses to one. Using this recursively permits to reduce the problem to the transport of a Dirac mass to a Dirac mass. This is the top-down part of the construction, i.e. the construction of the hierarchy of pivot points.

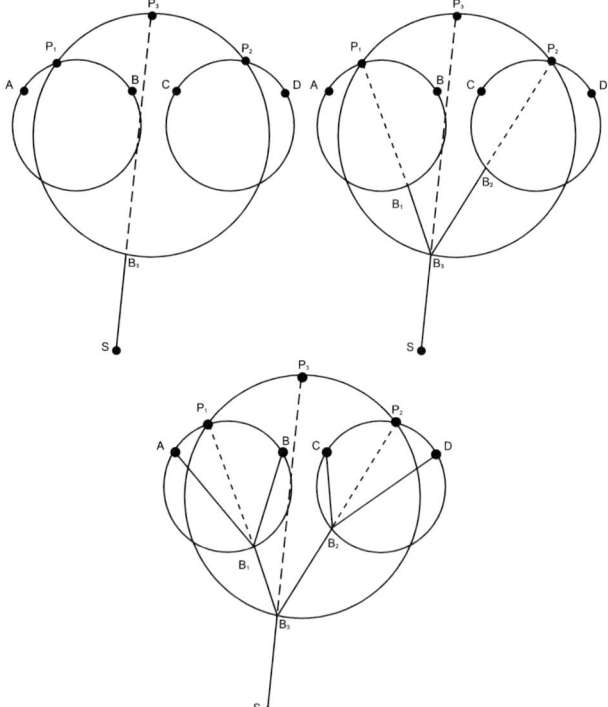

Fig. 12.9. The bottom-up part of the construction: connecting the source to the last pivot point permits to find the bifurcation point which is taken as the new source point for the two induced topologies.

$$\sup_{m\in]0,1]} g(m) = \begin{cases} 2^{2\alpha-1} - 1 & \text{if } \alpha > \frac{1}{2} \\ 0 & \text{if } 0 < \alpha \leq \frac{1}{2}. \end{cases}$$

Proof. The sign of g' is the same as the one of

$$\psi(\alpha) := (m+1)^{2\alpha} - 2(m+1)^{2\alpha-1} - m^{2\alpha} + 1.$$

We notice further that $\psi(0) = \frac{m-1}{m+1} < 0$, $\psi(1/2) = \psi(1) = 0$ and that ψ is concave. Indeed,

$$\psi''(\alpha) = 4(m+1)^{2\alpha} \ln(m+1)^2 - 8(m+1)^{2\alpha-1} \ln(m+1)^2 - 4m^{2\alpha} \ln(m)^2$$

$$= 4(m+1)^{2\alpha-1}(m-1) \ln(m+1)^2 - 4m^{2\alpha} \ln(m)^2$$

$$\leq 0.$$

The last inequality results from the fact that $m \leq 1$. Thus, $\psi(\alpha) \geq 0$ for $1 > \alpha > \frac{1}{2}$ and g' is non-negative so that g is nondecreasing. Similarly, $\psi(\alpha) \leq 0$ for $\alpha < \frac{1}{2}$ and g' is non-positive so that g is nondecreasing. The monotonicity of g permits to easily calculate the supremum,

$$\sup_{m\in]0,1]} g(m) = \begin{cases} g(1) = 2^{2\alpha-1} - 1 & \text{if } \frac{1}{2} \leq \alpha \leq 1 \\ \lim_{m\to 0} g(m) = 0 & \text{if } 0 < \alpha \leq \frac{1}{2}. \end{cases}$$

Lemma 12.15. *Let* $e_1 = BA_1$ *and* $e_2 = BA_2$ *be two oriented edges with length* r. *Let* $\boldsymbol{P}$ *be a traffic plan made of the two edges* e_1 *and* e_2 *with masses* m_1 *and* m_2. *If* $\boldsymbol{P}$ *is optimal, the angle* θ *between* e_1 *and* e_2 *is such that* $\cos(\theta) \leq 2^{2\alpha-1} - 1$ *for* $\frac{1}{2} < \alpha \leq 1$ *and* $\cos(\theta) \leq 0$ *for* $0 < \alpha \leq \frac{1}{2}$.

Proof. Because of Lemma 12.2 and Lemma 12.14,

$$\cos(\theta) \leq \sup_{m_1,m_2\in[0,1]} \frac{(m_1+m_2)^{2\alpha} - m_1^{2\alpha} - m_2^{2\alpha}}{2m_1^\alpha m_2^\alpha}$$

$$= \begin{cases} 2^{2\alpha-1} - 1 & \text{if } \frac{1}{2} < \alpha \leq 1 \\ 0 & \text{if } 0 < \alpha \leq \frac{1}{2}. \end{cases}$$

Lemma 12.16. *Let* $e^+ = A^+ B$ *and* $e^- = BA^-$ *be two oriented edges with length* r. *Let* $\boldsymbol{P}$ *be a traffic plan made of the two edges* e^+ *and* e^- *with masses* m *and* m', $m \geq m'$. *If* $\boldsymbol{P}$ *is optimal, the angle* θ *between* e^+ *and* e^- *is such that* $\cos(\theta) \leq (\frac{m}{m'} - 1)^\alpha - (\frac{m}{m'})^\alpha$. *In particular,* θ *is strictly superior to* $\frac{\pi}{2}$.

Proof. Without loss of generality, we can suppose that $m \geq m'$. Let B_ϵ be the point on segment A^+B at a distance ϵ of B. Let us consider the traffic plan $\boldsymbol{P}_\epsilon$ made of the edges $(A^+B_\epsilon, m), (B_\epsilon A^-, m')$ and $(B_\epsilon B, m - m')$. Let us denote

$$\delta(\epsilon) = E^\alpha(\boldsymbol{P}) - E^\alpha(\boldsymbol{P}_\epsilon)$$
$$= m^\alpha + m'^\alpha - (m^\alpha(1 - \epsilon) + (m - m')^\alpha \epsilon + m'^\alpha \sqrt{1 + \epsilon^2 - 2\epsilon \cos(\theta)}).$$

Since the traffic plan $\boldsymbol{P}_\epsilon$ has the same transference plan as $\boldsymbol{P}$ and $\boldsymbol{P}$ is optimal, $E^\alpha(\boldsymbol{P}_\epsilon) \geq E^\alpha(\boldsymbol{P})$, i.e. $\delta(\epsilon) \leq 0$. Thus $\delta'(0) \leq 0$, i.e. $\cos(\theta) \leq (\frac{m}{m'} - 1)^\alpha - (\frac{m}{m'})^\alpha$. In particular, $\cos(\theta) < 0$ so that $\theta > \frac{\pi}{2}$.

Proposition 12.17. *Let $\alpha \leq 1/2$ and $\boldsymbol{P}$ be an optimal traffic plan of $\mathbb{R}^2$ with finite graph structure. A branching point of the graph not in the support of μ^+ and μ^- has an edge multiplicity less than or equal to 3.*

Proof. Let B be a bifurcation point with more than three edges at B. Let us consider the restriction of the traffic plan $\boldsymbol{P}$ to a small ball of radius r around B within wich B is the only bifurcation. This restricted traffic plan is optimal for the irrigation problem from ν^+ to ν^- where ν^+ and ν^- are atomic measures on the circle $C(B, r)$. Let us denote by L^- and L^+ respectively the set of edges connecting B to ν^- and ν^+. All subtraffic plans made of two edges of L^+ or two edges of L^- are optimal. Depending on the flow direction in B, we may apply either Lemma 12.15 or Lemma 12.16, in the case $\alpha \leq \frac{1}{2}$, to conclude that the angle θ between two edges e and e' satisfies $\cos(\theta) \leq 0$. In any case, the angle between e and e' is greater or equal than $\frac{\pi}{2}$. This fact together with Lemma 12.16 implies that $\text{card}(L^+ \cup L^-) \leq 3$. Indeed, if $\text{card}(L^+ \cup L^-) \geq 4$, we could extract four edges from $L^+ \cup L^-$ forming angles greater than $\frac{\pi}{2}$. Since not all edges are in the same set L^+, or L^-, one of the angles is strictly greater than $\frac{\pi}{2}$ because of Lemma 12.16. Thus, there is no room for more than three edges in $L^+ \cup L^-$.

Remark 12.18. There is a very quick and geometric argument to prove that no "Ψ" shape can occur for an optimal traffic plan when $\alpha \leq \frac{1}{2}$. It is illustrated by Figure 12.10 and Figure 12.11. The argument is the following. Let us suppose that a Ψ shape is optimal and denote B the bifurcation point. In particular the subtraffic plan made of edges BA_1 and BA_2 is optimal so that B lies within the disk D_1 defined by the equiangle arc of Proposition 12.10. In the same way, the subtraffic plan made of edges BA_2 and BA_3 is optimal so that B lies within the disk D_2 defined by the equiangle arc. For $\alpha \leq \frac{1}{2}$, $D_1 \cap D_2 = \emptyset$ so that we obtain a contradiction.

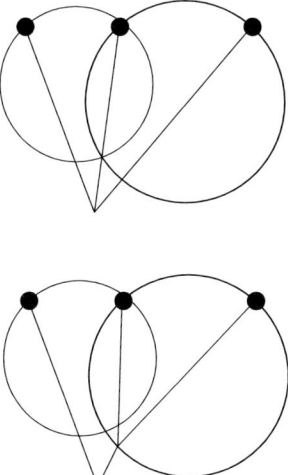

Fig. 12.10. If there is a triple point outside an equiangle arc, then a "ψ" structure can be improved as illustrated, thanks to Proposition 12.10.

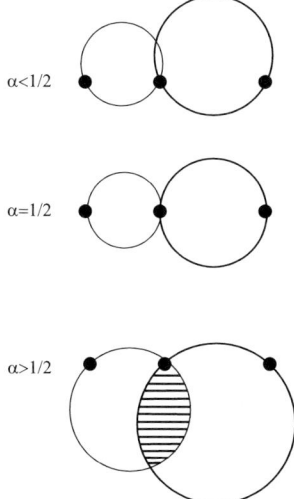

Fig. 12.11. As illustrated by Figure 12.10, a triple point of an optimal traffic plan has to lie within the two corresponding equiangle circles. In the case $\alpha \leq 1/2$, the intersection of these two disks is empty (both figures at the top). In the case $1 > \alpha > 1/2$, the intersection is not empty so that we cannot conclude immediately.

13 Dirac to Lebesgue Segment: A Case Study

In this chapter, we investigate the structure of an optimal traffic plan irrigating the Lebesgue measure on the segment from a single source. In the case of Monge-Kantorovich transport, as illustrated by Figure 13.1, an optimal traffic plan is totally spread in the sense that fibers connect every point of the segment with the source. If $\alpha = 0$, which corresponds to the problem of Steiner, an optimal traffic plan is such that all the mass is first conveyed to some point of the segment and then sent onto the whole segment. What is the shape of an optimal traffic plan for an intermediate $\alpha \in (0,1)$? This is the question we explore in this chapter, both from an analytical and experimental point of view. In the first section, we first prove in Lemma 13.6 that an optimal traffic plan cannot be like the one for $\alpha = 0$, i.e. with no bifurcation away from the segment(configuration that we call $T-$structure). Indeed, we prove by a perturbation argument that there exists a traffic plan with a $Y-$structure that has a lower cost than the $T-$ structure one. This implies that there have to be an infinite number of bifurcations (see Corollary 13.11). Indeed, if there were a finite number of bifurcations, we could extract an optimal traffic plan with $T-$structure (see Figure 13.5). With the same perturbation technique, we finally prove in Proposition 13.12 that whenever the traffic plan conveys some positive mass on the segment, the path that accomplishes this transportation has to be tangent to the segment. In the second section we discuss different algorithms to obtain local or global optima for the problem of transporting a Dirac mass to a discretized version of the Lebesgue segment, i.e. a finite atomic measure made of n Dirac masses with weight $\frac{1}{n}$. We first introduce a notation that permits to describe the different possible topologies of the reasonable trees from a source to n masses. The Gilbert-Steiner recursive procedure described in Chapter 12, permits to construct a local minimum (if it exists) with a prescribed topology. However, an exhaustive search through all the possible topologies of course leads to a combinatorial explosion. Thus, to rapidly obtain local optima with low cost, if not global ones, we have to derive some efficient heuristics. We describe in particular a multiscale approach that permits to quickly obtain a good local optimum, that we afterwards improve by perturbation. The presentation and main results here follows closely Bernot's PhD [10].

Let μ^- be the Lebesgue measure on the segment $[0,1] \times \{0\}$ and $\mu^+ = \delta_S$ the Dirac mass at the point S. In the following, we discuss the structure of an optimal traffic plan from μ^+ to μ^-.

M. Bernot et al., *Optimal Transportation Networks*. Lecture Notes in Mathematics 1955.
© Springer-Verlag Berlin Heidelberg 2009

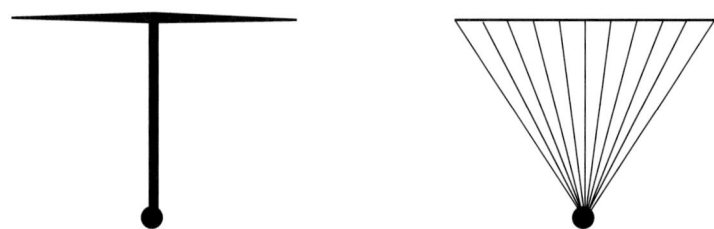

Fig. 13.1. In the case $\alpha = 0$ (resp. $\alpha = 1$), the irrigation problem amounts to find a structure connecting μ^+ and μ^- with minimal total length (resp. minimal average length of fibers). These optimal structures are respectively represented on the left-hand side and rigth-hand side of the figure. The question we try to answer in this chapter is: what is the structure of an optimal for $0 < \alpha < 1$?

13.1 Analytical Results

13.1.1 The Case of a Source Aligned with the Segment

Definition 13.1. *Let $\gamma : [0,1] \to X$ be a 1-Lipschitz curve parameterized by its arc-length, μ^- the uniform probability measure on $\gamma([0,1])$ and $\mu^+ = \delta_{\gamma(0)}$. We say that a traffic plan P from μ^+ to μ^- dissipates its mass along the curve γ if the speed normalization of P can be parameterized by $\chi(\omega, t) = \gamma(\min(\omega, t))$.*

In other words, we say that a traffic plan dissipates its mass along γ if it behaves as a bucket full of sand with a hole at the bottom and moving with a constant speed. We now determine an optimal traffic plan in the case where $S \in \mathbb{R} \times \{0\}$. If S is not on the segment, an optimal structure first transport the mass to the nearest tip of the segment then dissipates the mass along the segment. If S is on the segment, then an optimal traffic plan is the union of two traffic plans dissipating respectively on the left and right part of the segment separated by S.

Lemma 13.2. *Let $S = (s,0) \in \mathbb{R} \times \{0\}$ and P be an optimal traffic plan from μ^+ to μ^-. If $s \in]0,1[$ then P is the union of a traffic plan dissipating a mass s on $[0,s]$ and a traffic plan dissipating a mass $1 - s$ on $[s,1]$. If $s < 0$ then P is the concatenation of a traffic plan transporting μ^+ onto $\delta_{(0,0)}$ and a traffic plan dissipating along $[0,1]$. If $s > 1$ then P is the concatenation of a traffic plan transporting μ^+ onto $\delta_{(1,0)}$ and a traffic plan dissipating along $[0,1]$. In particular, if $S = (0,0)$, $E^\alpha(P) = \frac{1}{\alpha+1}$.*

Proof. The convex envelop property 5.15 tells that the support of P is in the line containing the segment. Because of the single path property, the mass has to dissipate uniformly along the fibers. Thus, $E^\alpha(P) = \int_0^1 x^\alpha dx = \frac{1}{\alpha+1}$.

Remark 13.3. A traffic plan $\boldsymbol{P}$ that dissipates a mass m onto a segment with length m is such that $E^\alpha(\boldsymbol{P}) = \frac{m^{\alpha+1}}{\alpha+1}$. Indeed, Lemma 13.2 ensures that the cost for mass 1 is $\frac{1}{\alpha+1}$. The case of a mass m is deduced from a scaling of the mass and of the length of the irrigated segment.

13.2 A "*T* Structure" is not Optimal

Definition 13.4. *We assume that $S \notin \mathbb{R} \times \{0\}$. Let $s \in [0,1]$ and δ_B the Dirac mass located at $B = (s,0)$. To every $s \in [0,1]$, we associate $\boldsymbol{P}_B$ the traffic plan obtained as the concatenation of the optimal traffic plan from μ^+ to δ_B and the optimal traffic plan from δ_B to μ^-. We say that such a traffic plan has a T structure.*

Remark 13.5. Notice that because of Lemma 13.2, a T structure is the concatenation of the oriented segment SB with mass 1 and the dissipation of a mass s and $1 - s$ from B to $[0, s]$ and $[s, 1]$ respectively. Thus, using Remark 13.3, the cost of a T structure $\boldsymbol{P}_B$ with branching point $B = (s, 0)$ is $E^\alpha(\boldsymbol{P}_B) = |SB| + \frac{1}{\alpha+1}(s^{\alpha+1} + (1-s)^{\alpha+1})$.

Lemma 13.6. *A traffic plan with T structure irrigating the Legesgue measure on the segment $[0, 1] \times \{0\}$ is not optimal.*

Proof. Let $\boldsymbol{P}_T$ be the T structure associated to $s \in [0, 1]$. Let us prove that it is possible to find a Y structure more efficient than the T one.
　　We first consider the case where $s \notin \{0, 1\}$. Let us define $\tilde{B}$ to be the point located at a distance ε from $B = (s, 0)$ and lying on the segment SB, B^- the point $(s - x, 0)$ and B^+ the point $(s + x, 0)$ where x is to be determined. We define the Y structure competitor (see Figure 13.2) as $\boldsymbol{P}_Y$, the concatenation of $S\tilde{B}$ with mass 1 with the union of the two T structures $\tilde{B}B^-$ with mass s and $\tilde{B}B^+$ with mass $1 - s$. Let $a = \|\tilde{B}B^-\|$ and $b = \|\tilde{B}B^+\|$. Notice that we have $a = \sqrt{\varepsilon^2 + x^2 + 2\varepsilon x \cos(\theta)}$ and $b = \sqrt{\varepsilon^2 + x^2 - 2\varepsilon x \cos(\theta)}$ where θ is the angle formed by the segments SB and OB where $O = (0, 0)$. Notice that since S is not aligned with the segment, we have $\theta \notin \pi\mathbb{Z}$.
　　The cost of this Y structure can be written as

$$E^\alpha(\boldsymbol{P}_Y) = \|S\tilde{B}\| + \phi(\varepsilon, x) \tag{13.1}$$

where

$$\phi(\varepsilon, x) = s^\alpha a + (1-s)^\alpha b + \frac{1}{\alpha+1}(2x^{\alpha+1} + (s-x)^{\alpha+1} + (1-s-x)^{\alpha+1}).$$

By Remark 13.5, the cost of $\boldsymbol{P}_T$ is $E^\alpha(\boldsymbol{P}_T) = \|S\tilde{B}\| + u(\varepsilon)$, where

$$u(\varepsilon) = \varepsilon + \frac{1}{\alpha+1}(s^{\alpha+1} + (1-s)^{\alpha+1}).$$

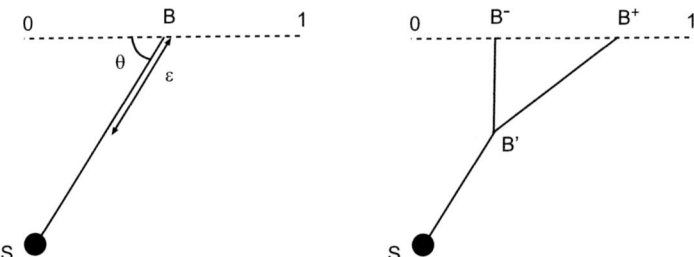

Fig. 13.2. A T structure with branching point B and a perturbation of it with Y structure.

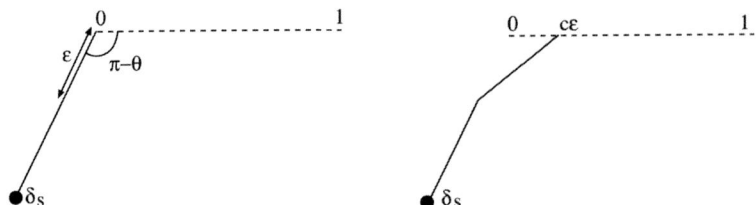

Fig. 13.3. A degenerate T structure and a perturbation of it.

Let us denote $v(\varepsilon) = \phi(\varepsilon, c\varepsilon)$. Notice that $v(0) = u(0)$ and $u'(0) = 1$. Thus it is sufficient to show that for some suitable $c > 0$, $v'(0) < 1$ so that $v(\varepsilon) < u(\varepsilon)$ for ε small enough. Let us calculate the derivative of v at the origin,

$$v'(0) = \left(\sqrt{c^2 + 1 + 2c\cos(\theta)} s^\alpha + \sqrt{c^2 + 1 - 2c\cos(\theta)}(1-s)^\alpha \right)$$
$$-c(s^\alpha + (1-s)^\alpha).$$

For $c = 0$, $v'(0) = s^\alpha + (1-s)^\alpha$. For c near infinity, the asymptotic expansion of $v'(0)$ is

$$v'(0) = c\left(s^\alpha(1 + \frac{\cos(\theta)}{c} + O(\frac{1}{c^2})) + (1-s)^\alpha(1 - \frac{\cos(\theta)}{c} + O(\frac{1}{c^2})) \right)$$
$$-c(s^\alpha + (1-s)^\alpha)$$
$$= \cos(\theta)(s^\alpha - (1-s)^\alpha) + O(\frac{1}{c}).$$

Since $s \notin \{0,1\}$ and $v'(0)$ is continuous with respect to c, we deduce that $v'(0) < 1$ for c large enough.

Let us now prove that a T structure is not optimal in the case $s \in \{0,1\}$. By symmetry we consider only the case $s = 0$. Let us define $\tilde{B}$ and B^+ as in the case where $s \in]0,1[$. We consider the new competitor (see Figure 13.3) as being the concatenation of $S\tilde{B}$ with mass 1 and the T structure from $\tilde{B}$ to B^+. The cost of this traffic plan is again given by (13.1) where

$$\phi(\varepsilon, x) = \sqrt{\varepsilon^2 + x^2 + 2\varepsilon x \cos(\theta)} + \frac{1}{\alpha + 1}(x^{\alpha+1} + (1 - x)^{\alpha+1}).$$

If $v(\varepsilon) = \phi(\varepsilon, c\varepsilon)$, then $v(0) = u(0)$ and

$$v'(0) = \sqrt{1 + 2\cos(\theta) + c^2} - c,$$

so that $v'(0) = \cos(\theta) + O(\frac{1}{c})$ for c large enough. Since $u'(0) = 1$, the T structure with a junction at $s = 0$ or $s = 1$ is not optimal and we can find a better T structure with junction in $]0, 1[$.

13.3 The Boundary Behavior of an Optimal Solution

In all that section, we assume that $S = (S_1, S_2) \in \mathbb{R}^2$ with $S_2 < 0$ and consider χ to be an optimal traffic plan from $\mu^+ = \delta_S$ to μ^-, the Lebesgue measure on $[0, 1] \times \{0\}$.

Definition 13.7. *Let χ_0 be the connected component of χ in $\mathbb{R}^2 \setminus [0, 1] \times \{0\}$. We call the boundary measure of χ the measure ν irrigated by χ_0.*

Lemma 13.8. *Let ν be the boundary measure of χ. If ν contains a Dirac mass at some point $\mathbf{y}_0 \in [0, 1] \times \{0\}$ and $\chi_{\mathbf{y}_0}$ is the subtree of χ starting at $\mathbf{y}_0$, then $\chi_{\mathbf{y}_0}$ irrigates the Lebesgue measure on a subinterval $[y_0 - \Delta_l, y_0 + \Delta_r] \times \{0\}$ of $[0, 1] \times \{0\}$ where $\Delta_l, \Delta_r \geq 0$, $\Delta_l + \Delta_r > 0$.*

Proof. Recall that χ is a trunk tree. Assume that ν contains a Dirac mass at $\mathbf{y}_0 = (y_0, 0)$ and let $\chi_{\mathbf{y}_0}$ be the subtree starting at $\mathbf{y}_0$. Then $\chi_{\mathbf{y}_0}$ is an optimal trunk tree that irrigates a measure $\mu_{\mathbf{y}_0}$ supported on a subinterval $[y_0 - \Delta_l, y_0 + \Delta_r] \times \{0\}$ of $[0, 1] \times \{0\}$ with density $f(x)dx$, $f \geq 0$. Here $\Delta_l, \Delta_r \geq 0$ and both cannot be zero. Let x_0 be a density point of $f(x)$ such that $f(x_0) > 0$. If $x_0 > y_0$, then $|x|_\chi > 0$ for any $x \in [y_0, x_0)$ since a positive flux has to be sent to a neighborhood of x_0 through the segment $[y_0, x_0]$. Then no good fiber of χ other than those in $\chi_{\mathbf{y}_0}$ can arrive to x, since by Santambrogio's Corollary 2.1, the multiplicity of the final point of two good curves is zero. We argue in a similar way when $x_0 < y_0$, and we deduce that $\chi_{\mathbf{y}_0}$ irrigates the Lebesgue measure in $[y_0 - \Delta_l, y_0 + \Delta_r] \times \{0\}$.

Lemma 13.9. *Let ν be the boundary measure of χ such that ν contains a Dirac mass at some point $\mathbf{y}_0 \in [0, 1] \times \{0\}$ irrigating the Lebesgue measure on $[y_0 - \Delta_l, y_0 + \Delta_r] \times \{0\}$, $\Delta_l, \Delta_r \geq 0$, $\Delta_l + \Delta_r > 0$. Let Γ be the unique path going from the source S to $\mathbf{y}_0$. If $\Delta_r > 0$ (resp. $\Delta_l > 0$), then there are at most a finite number of bifurcations to the right (resp. to the left) of Γ near $\mathbf{y}_0$.*

Proof. Assume that $\Delta_r > 0$ and that there are at most a finite number of bifurcations to the right of Γ near $\mathbf{y}_0$. Notice that, by Lemma 13.8, they end at the right of $y_0 + \Delta_r$. Let us denote by $\mathbf{y}_i$ the points where Γ has a bifurcation to the right. Let $\chi_{\mathbf{y}_i}$ be the subtrees of χ that start on $\mathbf{y}_i$ (see Figure 13.4). They are optimal and disjoint subtrees and they irrigate a measure $\mu_{\mathbf{y}_i}$ with support on an interval of $[0,1] \times \{0\}$ to the right of $\mathbf{y}_0 + \Delta_r$. Since $\cup_i \chi_{\mathbf{y}_i}$ is optimal, it has to be finite by Lemma 8.9.

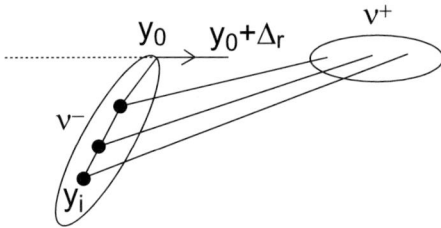

Fig. 13.4. See proof of Lemma 13.9. Let us consider a path connecting the source to a Dirac mass in $[0,1] \times \{0\}$ which dissipates on the right and has an infinite number of bifurcations to the right. The traffic plan defined by the union of subtrees bifurcating to the right irrigates a measure located away of $\mathbf{y}_0$. By Lemma 8.9, since this union is optimal, it has to be finite.

Proposition 13.10. *Let ν be the boundary measure of χ such that ν contains a Dirac mass at some point $\mathbf{y}_0 \in [0,1] \times \{0\}$. Let $\chi_{\mathbf{y}_0}$ be the subtree of χ starting at $\mathbf{y}_0$. Then $\chi_{\mathbf{y}_0}$ irrigates the Lebesgue measure on some $[y_0 - \Delta_l, y_0 + \Delta_r]$ where $\Delta_l, \Delta_r \geq 0$, $\Delta_l \Delta_r = 0$ and $\Delta_l + \Delta_r > 0$. Moreover, let Γ be the unique path going from the source S to $\mathbf{y}_0$. If $\Delta_r > 0$ (resp. $\Delta_l > 0$), then Γ has an infinite number of bifurcation on the left (resp. right) and a finite number of bifurcation on the right (resp. left).*

Proof. Lemma 13.8 asserts that $\chi_{\mathbf{y}_0}$ irrigates the Lebesgue measure on $[y_0 - \Delta_l, y_0 + \Delta_r]$. Let Γ be the unique path going from the source S to $\mathbf{y}_0$ and assume that $\Delta_r > 0$ and $\Delta_l > 0$. Then Lemma 13.9 ensures that there is a finite number of bifurcations near $\mathbf{y}_0$. Thus, as illustrated by Figure 13.5, there is a point $\mathbf{y}_1$ lying on Γ and such that $\chi_{\mathbf{y}_1}$ (which is optimal as a subtree of the optimal traffic plan χ) has a T structure. This is a contradiction with Lemma 13.6 stating that a T structure cannot be optimal. Thus, $\Delta_l = 0$ or $\Delta_r = 0$. Both of these quantities cannot be null simultaneously because we are irrigating the Lebesgue measure on the segment and no Dirac mass at any point $\mathbf{y}_0$. Thus we have $\Delta_l \Delta_r = 0$ and $\Delta_l + \Delta_r > 0$. Without loss of generality, let us consider the case when $\Delta_r > 0$. Then Lemma 13.9 asserts that there is a finite number of bifurcations of Γ on the right. If there is a finite number of bifurcations of Γ on the left, then there is a last bifurcation point $\mathbf{y} \neq \mathbf{y}_0$. The concatenation of the segment from $\mathbf{y}$ to $\mathbf{y}_0$ and $\chi_{\mathbf{y}_0}$ is an optimal traffic plan with a degenerated T structure and this contradicts Lemma 13.6. We conclude that Γ has an infinite number of bifurcations on the left.

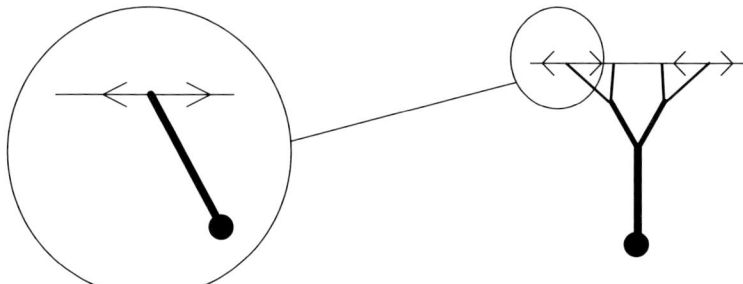

Fig. 13.5. See proof of Proposition 13.10. Let ν be the boundary measure of P. If ν has a Dirac mass dissipating mass both sides, we can consider an optimal subtree of P that has a T structure. This is a contradiction with Lemma 13.6. Dissipation cannot take place both sides.

Corollary 13.11. *Let P be an optimal traffic plan from $\mu^+ = \delta_S$ to μ^-, the Lebesgue measure on $[0,1] \times \{0\}$. Let χ_0 be the connected component of χ in $\mathbb{R}^2 \setminus [0,1] \times \{0\}$. If ν is the measure irrigated by χ_0, then ν is not finite atomic.*

In other words, an optimum is not finite graph plus dissipation of mass along the segment structure.

Proof. If the boundary measure ν is a finite atomic measure, Proposition 13.10 ensures that a path from the source to a Dirac mass of ν has an infinite number of branching points away from μ^+ and μ^-. However, since χ_0 is optimal, and ν is supposed to be atomic, Lemma 9.1 asserts that χ_0 has the structure of a finite graph. In particular, it has a finite number of bifurcations away from μ^+ and μ^-. This contradiction proves that ν cannot be a finite atomic measure.

Proposition 13.12. *Let ν be the boundary measure of an optimal χ and suppose that ν contains a Dirac mass of weight m at some point $\mathbf{y}_0 \in [0,1] \times \{0\}$ with dissipation on the right. Let Γ be the unique path from S to $\mathbf{y}_0$. Let $\mathbf{y} \in \Gamma$, and let $\theta(\mathbf{y})$ be the angle formed by the segments $\mathbf{y}\mathbf{y}_0$ and $\mathbf{0}\mathbf{y}_0$. Then $\theta(\mathbf{y}) \to 0$ as $\mathbf{y} \to \mathbf{y}_0$.*

Proof. To fix ideas, let us assume that $S = (S_1, S_2)$ with $S_2 < 0$. Observe that $m\delta_{\mathbf{y}_0}$ is dissipated onto the interval $[y_0, y_0 + m]$ where $\mathbf{y}_0 = (y_0, 0)$. Assume that there is a sequence of points $\mathbf{y}_\varepsilon \to \mathbf{y}_0$ such that $\theta_\varepsilon := \theta(\mathbf{y}_\varepsilon) \geq \theta > 0$. Let $d_\varepsilon = \|\mathbf{y}_\varepsilon - \mathbf{y}_0\|$. Let $\mathbf{x}_\varepsilon \in [0,1] \times \{0\}$ be a point to the right of $\mathbf{y}_0$ such that $\|\mathbf{x}_\varepsilon - \mathbf{y}_0\| = d_\varepsilon$. Let us consider the segment $\mathbf{y}_\varepsilon \mathbf{x}_\varepsilon$ whose length is

$$d'_\varepsilon = d_\varepsilon \sqrt{2 + 2\cos\theta_\varepsilon}.$$

Let $f(s) = m + \rho(s)$ be the flow along Γ where s denotes the arclength of Γ. We take $f(0) = m$ to be the flow at $\mathbf{y}_0$ and $f(\mathcal{H}^1(\Gamma)) = 1$ to be the flow at the source point S. Thus, $\rho(s)$ is a non decreasing function such that $\rho(0) = 0$. Let us consider a competitor of χ as the traffic plan χ_ε obtained by deviating the flow m through the T structure consisting of the segment $\mathbf{y}_\varepsilon \mathbf{x}_\varepsilon$ and then dissipating the mass d_ε to the left and the mass $m - d_\varepsilon$ to the right (see Figure 13.6). We have,

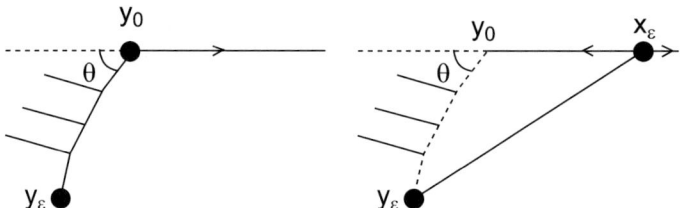

Fig. 13.6. See proof of Proposition 13.12. If the path connecting the source to a Dirac mass of the boundary measure has a limit angle $\theta > 0$, then we define a competitor χ_ε represented on the right-hand side of the figure that has lower cost than χ. Thus a path irrigating a Dirac mass of the boundary measure has to arrive tangentially on the segment.

$$E^\alpha(\chi) - E^\alpha(\chi_\varepsilon) = \int_{\Gamma_\varepsilon} [(m + \rho(s))^\alpha - \rho(s)^\alpha]\, ds - m^\alpha d_\varepsilon'$$
$$+ \frac{1}{\alpha + 1} \left(m^{\alpha+1} - d_\varepsilon^{\alpha+1} - (m - d_\varepsilon)^{\alpha+1}\right),$$

where Γ_ε is the arc of Γ from $\mathbf{y}_\varepsilon$ to $\mathbf{y}_0$. Since $\rho(s) \leq \rho(\mathcal{H}^1(\Gamma_\varepsilon))$ for $s \in [0, \mathcal{H}^1(\Gamma_\varepsilon)]$, we have,

$$\int_{\Gamma_\varepsilon} [(m + \rho(s))^\alpha - \rho(s)^\alpha]\, ds \geq (m^\alpha - o(1))\mathcal{H}^1(\Gamma_\varepsilon)$$
$$\geq m^\alpha d_\varepsilon + o(d_\varepsilon).$$

In addition,

$$-m^\alpha d_\varepsilon' + \frac{1}{\alpha + 1} \left(m^{\alpha+1} - d_\varepsilon^{\alpha+1} - (m - d_\varepsilon)^{\alpha+1}\right)$$
$$\geq -m^\alpha d_\varepsilon \sqrt{2 + 2\cos\theta_\varepsilon} + \frac{1}{\alpha + 1} \left(-d_\varepsilon^{\alpha+1} + (\alpha + 1)m^\alpha d_\varepsilon\right) + o(d_\varepsilon)$$
$$\geq -m^\alpha d_\varepsilon \sqrt{2 + 2\cos\theta_\varepsilon} + m^\alpha d_\varepsilon + o(d_\varepsilon).$$

Collecting both estimates, we obtain

$$E^\alpha(\chi) - E^\alpha(\chi_\varepsilon) \geq m^\alpha d_\varepsilon(2 - \sqrt{2 + 2\cos\theta}) + o(d_\varepsilon) > 0,$$

for $\varepsilon > 0$ small enough. This contradiction proves that $\theta(\mathbf{y}) \to 0$ as $\mathbf{y} \to \mathbf{y}_0$.

13.4 Can Fibers Move along the Segment in the Optimal Structure?

If $\alpha = 1$, the irrigation problem coincides with the Monge-Kantorovich one; then the boundary measure ν of an optimal traffic plan from a Dirac mass to the Lebesgue measure on the unit segment is precisely the Lebesgue measure on the segment. In the case $\alpha = 0$, ν consists of a single Dirac mass. By Corollary 13.11 we know that, for $0 < \alpha < 1$, ν cannot be a finite atomic measure. However ν may have Dirac masses (in which case dissipation can occur only in one direction at that Dirac mass). The questions we raise are the following :

- Are there Dirac masses in ν?
- Is ν purely made of Dirac masses?

In the authors point of view, the most appealing conjecture would be an affirmative answer to the last question (see Chapter 15).

13.5 Numerical Results

13.5.1 Coding of the Topology

Let $A = (a_i)_i$ be N points of the space. When the points $(a_i)_i$ are ordered on a line, it does not make sense to group first a_1 with a_3 and a_2 with a_4. No such mixing can occur in the case of an optimal structure, otherwise there would be a circuit which is impossible thanks to Proposition 7.8. Thus, we can restrict to "parenthesis" topologies, i.e. to topologies corresponding to all the possible ways to do the non-associative product $a_1...a_n$. We present here a convenient way to code for "parenthesis" topologies and to generate them all.

Definition 13.13. *All parenthesis topologies are recursively described by a list $[t_1, t_2, .., t_{n-1}]$. The coding works as follow: t_1 denotes the index of the first grouping so that we shrink a_{t_1}, a_{t_1+1} to a single formal point*

$$b_1 := (a_{t_1}, a_{t_1+1}).$$

Then $[t_2, ..., t_{n-1}]$ describes the topology, of $a_1, ..., a_{t_1-1}, b_1, a_{t_1+2}, ..., a_n$ (see Figure 13.7).

It is a classic combinatorial problem to enumerate these topologies.

Lemma 13.14. *The total number of topologies for N aligned points is the Catalan number $\frac{1}{n}\binom{2n-2}{n-1}$.*

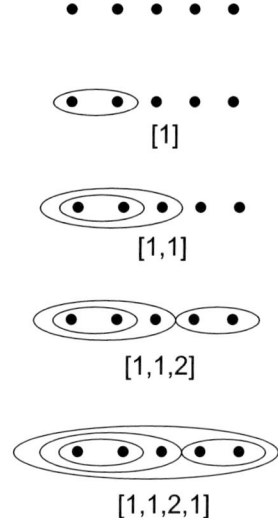

Fig. 13.7. The hierarchy of grouping is coded by a chain of numbers indicating the position of the successive merging.

13.5.2 Exhaustive Search

Let us briefly mention that the coding of topologies is particularly adapted to the pivot point algorithm since it permits a recursive description of the topology. Thus, in the case of few Dirac masses at the target measure, it is possible to proceed to an exhaustive search through all topologies. Figures 13.8 and 13.9 show for instance a full treatment of the irrigation of four aligned Dirac masses. In the case of many Dirac masses, such a treatment considering all cases is impossible. Figure 13.10 shows a topologically plausible solution for 64 aligned Dirac masses, when the exponent is close to 1. But it is just one of many possible solutions!

13.6 Heuristics for Topology Optimization

As it was said before, the irrigation problem can be divided into two optimization problems: the optimization of the topology, and the optimization of the locations of bifurcation points. The recursive construction presented in section 12.2 answers to the second optimization problem with an accurate construction along with an exhaustive search through all possible degeneracies of a topology. However, an exhaustive search through all topologies takes a lot of time and increasing the number of Dirac masses causes combinatorial explosion. Several heuristics can help in finding a reasonable topology within a reasonable time or in improving it. We present three of them:

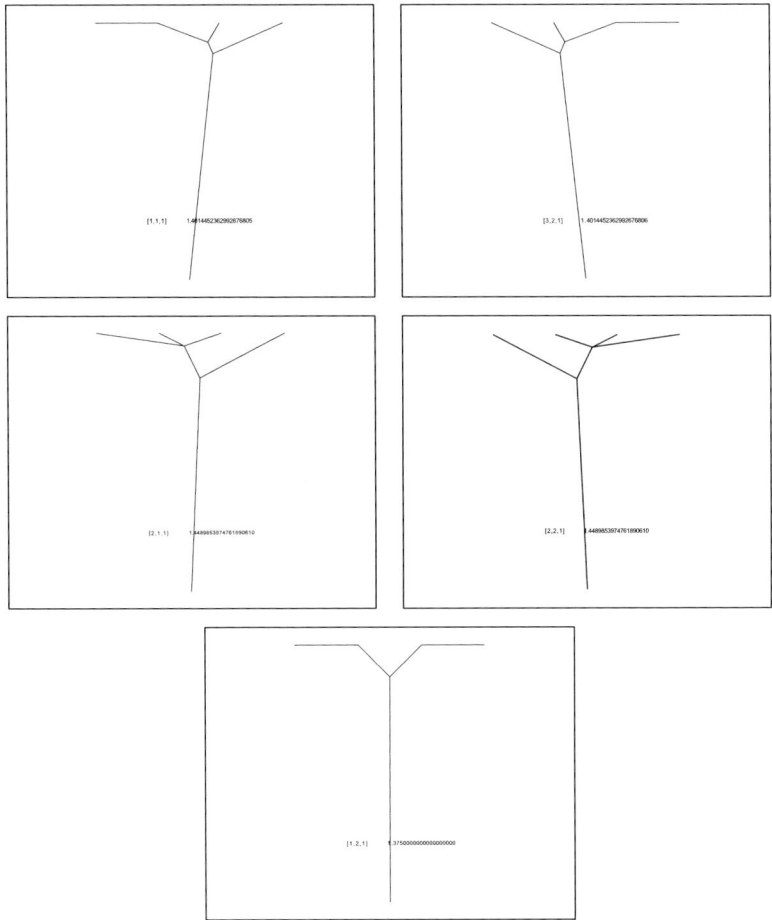

Fig. 13.8. All local optima associated to each topology for $\alpha = \frac{1}{2}$.

- The multiscale approach permits to find efficient topologies thanks to a compromise between accuracy of the resolution of the target measure and exhaustive search.
- The optimality of subtrees criterion looks if it is possible to improve some subtrees of the global structure.
- The perturbation method permits to move from a topology to another, allowing global improvement.

13.6.1 Multiscale Method

When the Dirac masses of the target measure are too numerous, the total number of possible topologies is much too big for the exhaustive exploration

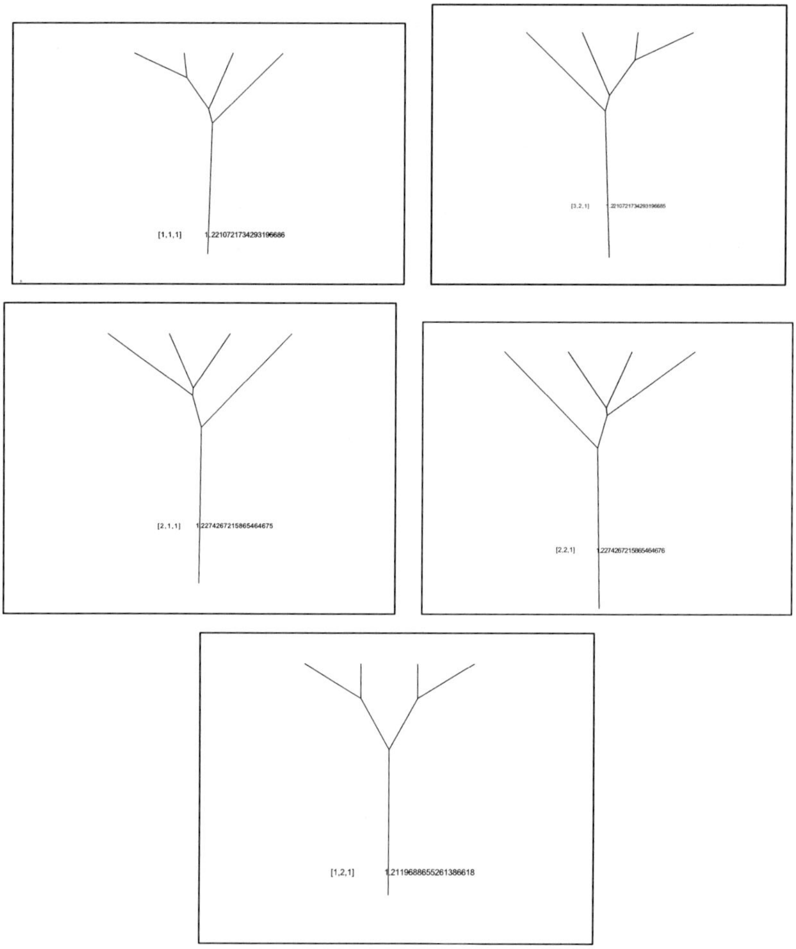

[1,1,1]   1.221072173429319686

[3,2,1]   2167217342931966885

[2,1,1]   1.227426721586546675

[2,2,1]   1.227426721586546676

[1,2,1]   1.211968865526138618

Fig. 13.9. All local optima associated to each topology for $\alpha = 0.8$.

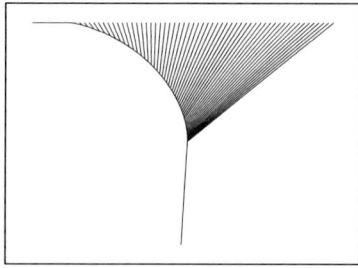

Fig. 13.10. The [1,1,...] topology for the irrigation of a 64-approximation of Lebesgue measure on the segment ($\alpha = 0.95$).

to take place. The multiscale approach permits to reduce the number of target points, and thus reduce the problem to a tractable one. The solution of this approximate problem gives hints on the structure of a good structure for the initial problem. These hints permit to reduce the initial problem to appropriate subtrees problems. The synthesis of all subtrees problems can then take place to obtain a reasonable (but not necessarily optimal) structure.

Let us illustrate how the multiscale approach works with $\mu^- = \lambda_{64}$ being the target measure (where λ_k denotes an approximation of the Lebesgue segment with k equal Dirac masses), μ^+ the source point at (0,-1) and $\alpha = 0.95$.

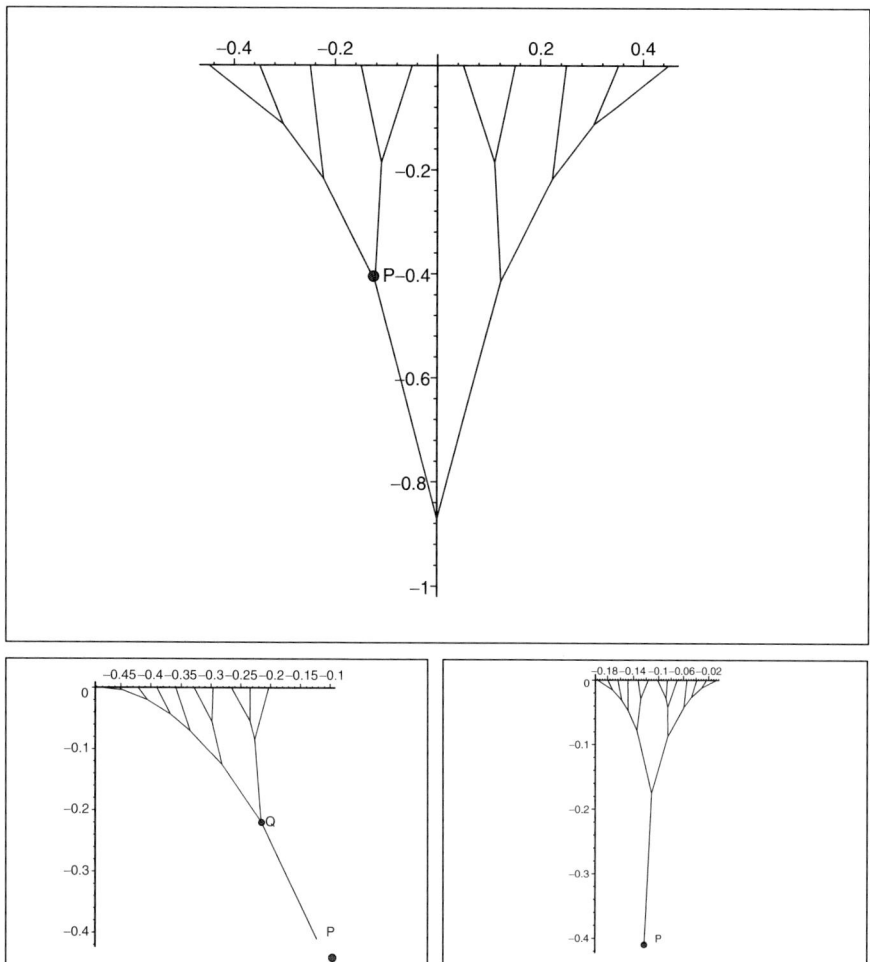

Fig. 13.11. At the top, T_{10} is transporting μ^- to $\mu^+ = \lambda_{10}$. The bifurcation point P induces two subtrees. The two figures at the bottom represent these two subtrees at a better resolution so that we can continue the multiscale optimization process.

- Best structure at a lower resolution: let us start by considering the optimal traffic plan transporting μ^- to $\mu^+ = \lambda_{10}$, we denote it by T_{10}. It is represented on Figure 13.11. This tree is symmetrical and because of the symmetry of the problem we shall look for a symmetrical solution.
- Two subtrees: we denote by P the second bifurcation point of T_{10}, it is located at (-0.123,-0.41). Two subtrees are starting from P, T_{10}^l on the left and T_{10}^r on the right.
- The range of the two subtrees: T_{10}^l irrigates target Dirac masses within $[-0.5, -0.2]$ and T_{10}^r irrigates target Dirac masses within $[-0.2, 0]$.
- Go back to the initial resolution: let μ_l^+ and μ_r^+ be respectively the sum of Dirac masses of λ_{64} located within $[-0.5, -0.2]$ and within $[-0.2, 0]$. Because of the previous point we bring back the initial problem to the one of finding efficient structures to transport P to μ_l^+ and P to μ_r^+.
- Iteration of the process: since μ_r^+ is made of 13 Dirac masses, we proceed to an exhaustive search of the optimal structure. Since μ_l^+ is made of 19 Dirac masses we apply the multiscale approach to this problem.
- Best structure at a lower resolution: we denote by ν_{10} an approximation of μ_l^+ made of 10 Dirac masses. The best traffic plan T_{10}^2 represented on Figure 13.11 and 13.12 has a bifurcation point Q located at $(-0.215, -0.22)$.
- The range of the two subtrees (see Figure 13.12): the two subtrees starting from Q have range $[-0.5, -0.28]$ and $[-0.28, -0.2]$. The corresponding measures at the initial resolution ν_l and ν_r are respectively made of 14 and 5 Dirac masses. The problem of finding the best structure from P to μ_l^+ thus reduces to the one of finding the best irrigation from Q to ν_l and ν_r. An exhaustive search can do this job.
- Recombination (see Figure 13.13): we decomposed λ_{64} as $\lambda_{64}^+ + \lambda_{64}^-$, respectively the Dirac masses on the right and on the left. The multiscale approach made us consider λ_{64}^- as $\lambda_{64}^- = \nu_l + \nu_r + \mu_r^+$. The recombination of optimal structures from Q to ν_l and ν_r gives an efficient structure T_l from P to μ_l^+. We can then combine it with the structure T_r that transports P to μ_r^+.

13.6.2 Optimality of Subtrees

Given an optimal structure T, a subtree is optimal for the problem it induces. That is to say, if we look at the two trees T_1 and T_2 (see Figure 13.14) starting at a bifurcation point P of an optimal structure, these two trees have to be optimal. Indeed, if it was not the case, there would be better trees T_1' and T_2' such that a combination of T_1' and T_2' would give a better structure than T. Thus, it is possible to improve some structures, only trying to improve subparts of it. For example, since a target measure with 10 Dirac masses is computationally tractable, we can test all subtrees irrigating less than 10 Dirac masses in order to improve a structure.

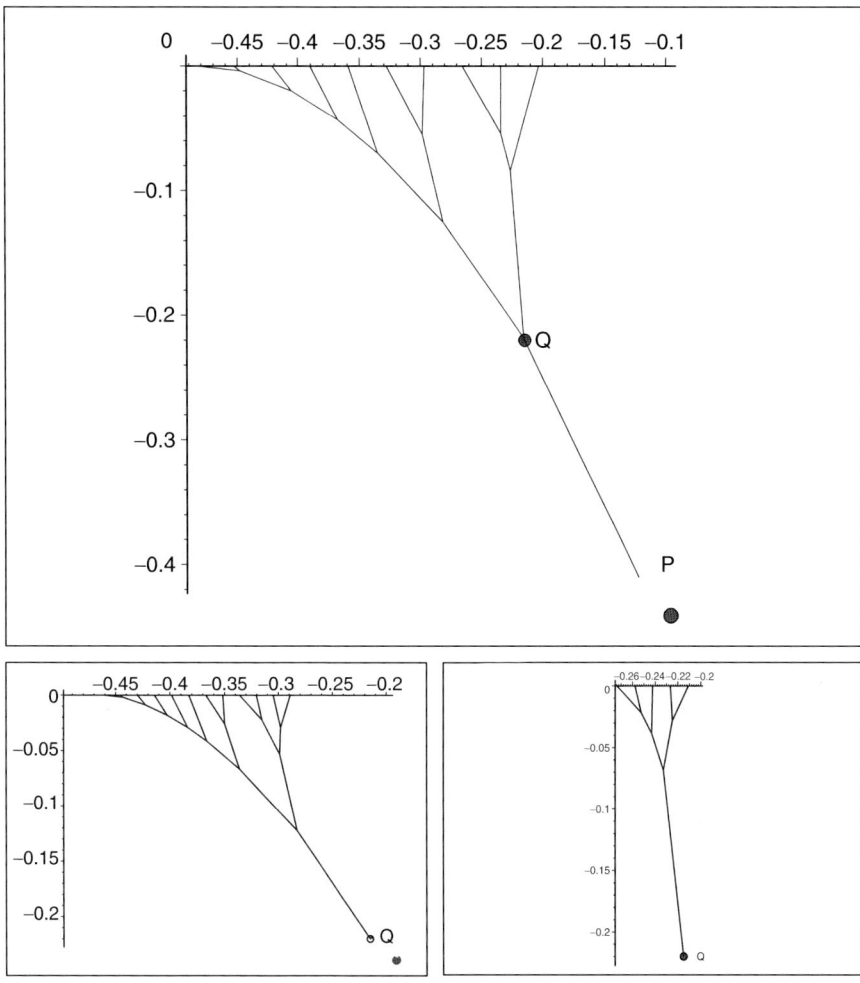

Fig. 13.12. The measure ν_{10} is an approximation of μ_r^+ that is made of 10 Dirac masses. The figure at the top represents T_{10}^2, the best traffic plan irrigating ν_{10} from P. The bifurcation point Q induces two subtrees, that we look at the initial resolution. The two figures at the bottom represent these two subtrees at the initial resolution.

13.6.3 Perturbation of the Topology

The third heuristic that permits to improve a given structure T consists in perturbing the topology of T. That is to say, given an edge e, we can define the topological neighborhood of (T, e) the set of topologies obtained through all possible perturbation of the edge e. In the case of parenthesis topologies, we reduce these perturbations to reasonable ones (see Figure 13.15).

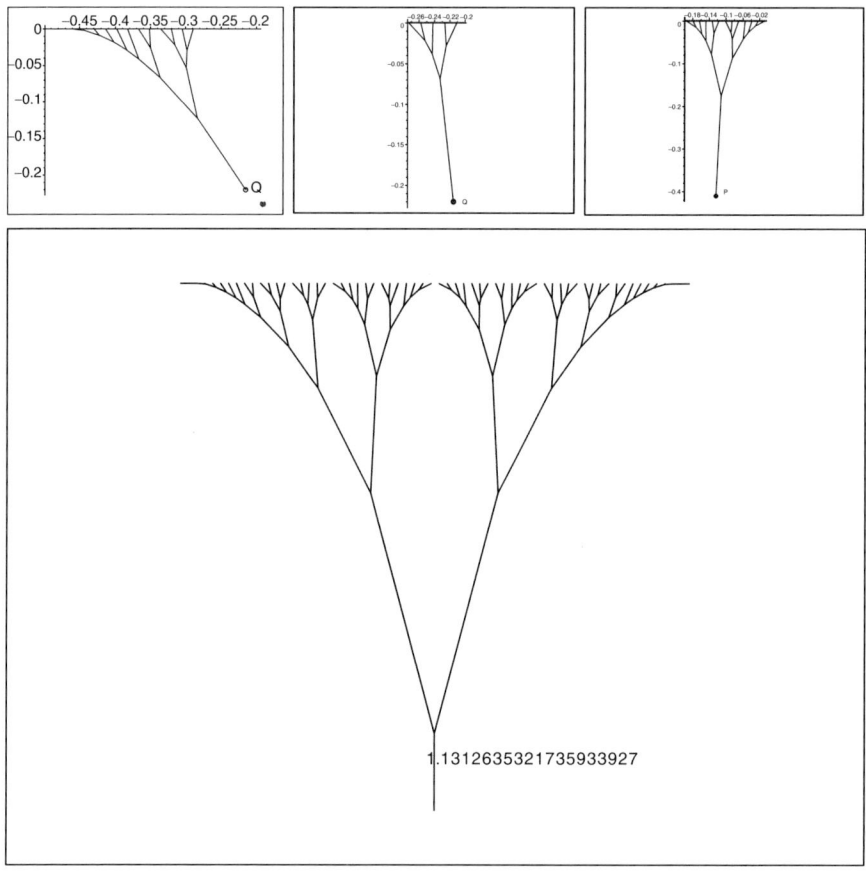

Fig. 13.13. We obtained efficient structures to transport a mass at Q to ν_l and ν_r and to transport a mass at P to μ_r^+. These three structures are represented at the top. The figure at the bottom represents the combination of these three structures that gives an efficient transport from the source point $(0, -1)$ to λ_{64}. Notice that this structure is better than the dyadic homogeneous one and has a cost 1.1312635 which is very close of the optimal one 1.1312238.

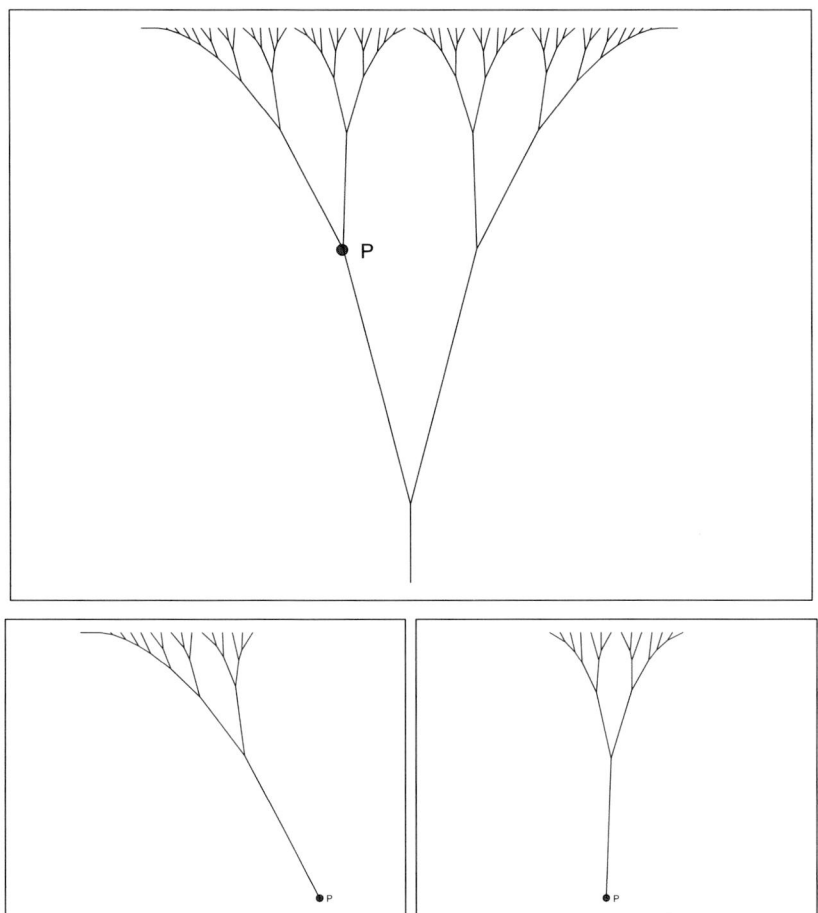

Fig. 13.14. If a tree is optimal, then all of its subtrees also have to be optimal. For instance the two subtrees starting from P are optimal in this case. This tells that we can't improve the initial structure with the optimality of subtrees criterion.

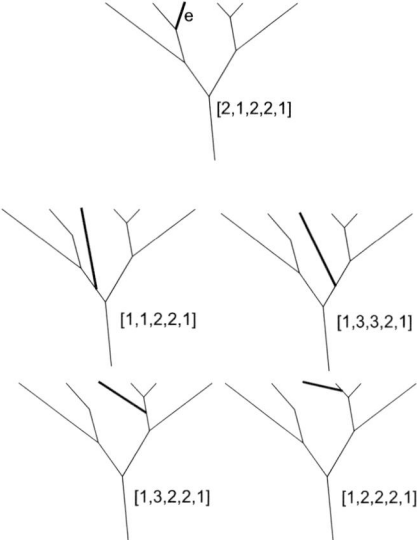

Fig. 13.15. The different possible topological perturbations associated to the edge e.

14 Application: Embedded Irrigation Networks

This chapter is devoted to a discussion of physical models of lungs, blood circulation and other natural or artificial networks irrigating a volume from a source. We shall address here the main objection to the model treated throughout these notes: they do not consider the fact that irrigation networks are embedded in a body. We have reduced the vessels or roads or tubes to infinitely thin paths, but they must in fact be thick enough. The Gilbert model reduces the three parameters of a tube, namely its length, diameter, and flow to only two, the length and the flow. Yet once we fix the flow to be conveyed in a tube we can make the tube very thick so that the flow becomes slow and has little kinetic energy. If instead the tube is very thin the fluid velocity must be large and therefore the dissipated power high. Thus, embedded models have apparently one more degree of freedom, the diameter of the tubes and this parameter changes the energy. By making the tube diameters tend to infinity we can even let the dissipated energy tend to zero. Of course embedded irrigation circuits cannot occupy the whole space and their volume must be constrained, unless we only grow lentils. Section 14.1 describes the physical principles and parameters of tube models. Section 14.2 shows that under a volume constraint the tube diameter can be eliminated in the expression of the dissipated power. This elimination is done by simple Lagrangian calculus and shows the dissipated power to be a Gilbert energy. The expression of this energy also leads to the interesting conclusion that, under the Poiseuille's law of laminar flow in the tubes, the α exponent is simply critical. One has $\alpha = 1 - \frac{1}{3}$ which was proven to be the largest exponent such that it is not possible to irrigate a positive measure set. This fact explains why all physical irrigation models consider networks whose capillaries have a minimal prefixed size. Otherwise, the energy of an irrigation circuit irrigating a whole volume is infinite. Section 14.3 considers embedded and not necessarily optimal irrigation patterns and gives a quick and more general proof that for any embedded pattern the volume and the energy cannot be simultaneously finite under Poiseuille's law. This chapter is an adaptation and simplification of [27], [12].

14.1 Irrigation Networks made of Tubes

We have seen that the function of many natural or artificial irrigation or drainage systems is to connect by a fluid flow a finite size volume to a source. A space filling hierarchical branching pattern is obviously required

M. Bernot et al., *Optimal Transportation Networks*. Lecture Notes in Mathematics 1955.
© Springer-Verlag Berlin Heidelberg 2009

and observed. Under the Gilbert-Steiner model we have proven several existence and regularity properties for such optimal patterns. The resulting irrigation circuitry is a tree of tubes branching from a source and going as close as possible to any point of the irrigated volume. In this chapter we go back to the physical irrigation models. The following principles have been proposed in this literature to characterize irrigation patterns :

(SF) Space filling requirement: the network supplies uniformly an entire volume of the organism.

(K) Kirchhoff's law at branching (conservation of fluid mass).

(W) Energy minimization: the biological networks have evolved to minimize energy dissipation.

(MSU) Minimal size unit: the size of the final branches of the network is a size-invariant unit.

These principles are considered basic principles in all presentations of irrigation circuits [88], [89], [90], [8]. In the case of irrigation or drainage networks, the energy criterion aims at a reduction of the overall resistance of the system, or, equivalently, to a minimization of the dissipated power.

In the mentioned papers several additional assumptions are usually made to derive conclusions from this set of principles which can be summarized in:

(H) Homogeneous tree: the irrigation system is assumed to be a tree made of tubes, fully homogeneous in lengths and sections.

Let us describe in some detail this homogeneous framework and its consequences. We denote by $k \in \mathbb{N}$, $k \leq k_{max}(\leq \infty)$ the branching level in the tree. The tubes at the final level k_{max} will be called the capillaries. There is a single tube at level 0, and N_k tubes at level k. By (H), at each level k all tubes are equal and are described by the same parameters: l_k, r_k, φ_k, namely the common value of their length, radius and flow. We shall also use the variable $s_k = r_k^2$ which is proportional to the area of the constant section of the tube. With these variables, the power dissipated by the irrigation network is expressed as

$$W = \sum_{k=1}^{k_{max}} N_k l_k s_k^{-\beta} \varphi_k^2. \tag{14.1}$$

Although we treat β as a free parameter, Poiseuille's law states that for all newtonian fluids in laminar mode, $\beta = 2$. The homogeneity of the irrigation tree can be rendered still more specific by imposing the realistic

(CB) Constant branching: $\frac{N_{k+1}}{N_k} = \nu = constant.$

The space filling requirement can be formalized in a rough way by stating that the k-th tube irrigates a volume proportional to l_k^3. This is a possible interpretation of (SF) which we shall call $(SF1)$. So we can summarize as a set of equations the constraints usually proposed for homogeneous trees

(H) Homogeneous tree with unknown k, l_k, r_k, φ_k, $k \leq k_{max}$.
(K) Kirchhoff $N_k \varphi_k = constant$.
(SF1) Space filling $N_k l_k^3 = constant$.
(MSU) Minimal size capillaries: $k_{max} < \infty$.
(CB) Constant branching (optional), $\frac{N_{k+1}}{N_k} = \nu = constant$.

The aim of this set of assumptions was in [88], [89], [90] to prove that the network has a fractal-like structure with self-similar properties. In the mentioned papers, it is claimed that the minimization of the energy (14.1) with prescribed volume $\sum_k N_k l_k r_k^2 = V$ leads to self-similar properties, namely constant ratios $\frac{l_{k+1}}{l_k} = constant$, $\frac{r_{k+1}}{r_k} = constant$, so that also $\frac{l_k}{r_k} = constant$, namely the tubes have a scale invariant shape and all quantities scale as powers of n. Actually, such results were not proven, the main focus of the mentioned papers being rather to discuss scaling laws in animal metabolism. A mathematically more comprehensive study of the consequences of the above mentioned axioms is given in [32] where the correct consequences are drawn.

14.1.1 Anticipating some Conclusions

The question arises of whether the four basic principles (SF), (K), (W) and (MSU) are compatible or not. One of the main outcomes of the next section will be to eliminate (MSU), that is, the minimal size constraint for the capillaries. The (MSU) assumption, essential in the above mentioned physical models, was simply written as $k_{max} < \infty$ and forbids infinite branching. It also actually excludes a volume direct irrigation and only permits any point of a volume to be "close enough" to a capillary. We shall prove that the only obstruction to infinite trees is the infinite resistance of such circuits. We assume without loss of generality that Poiseuille's law holds throughout the circuit: it is generally acknowledged that this law is valid in all biological circuits, at least for the smaller tubes [61]. The Poiseuille's law states that for fluids, the resistance $R(s) = Cs^{-\beta}$ of a tube with section s scales as the inverse second power of s. We shall prove that if $\beta \geq 2$, then $W = +\infty$ for any set of tubes obeying (K) and irrigating a positive volume (SF) (see Theorem 14.4 for a more precise statement). If we instead consider $R(s) = Cs^{-\beta}$ with $\beta < 2$, then infinitesimal circuits are possible. The power $\beta = 2$ is the limit exponent.
    Let us address the compatibility of (K), (W), (SF), (H) and (CB). Since Section 14.2 proves that minima of W actually are minima for a Gilbert energy, all of the study in the book applies. This study proves that (K), (W) and (SF) are compatible, provided $\alpha > 1 - \frac{1}{N}$, namely $\beta < 2$ in the case of Poiseuille's law. We have also proven (interior regularity) that the network looks like a tree of tubes away from the irrigated measures. Yet in the case of a volume irrigation network the tubes and the irrigated volume are interlaced and the "tree of tubes" model is too simplistic. The homogeneity assumption (H) is also clearly wrong as shown in the first numerical studies.

Finally (CB) is conjectural and we refer to Chapter 15. The fact that the number of branches has a universal bound has been proven in Section 8.4.

14.2 Getting Back to the Gilbert Functional

Consider a network irrigating a probability measure or $\mathbb{R}^3$. This network is assumed to be made of a finite or countable set of tubes indexed by $k \in \mathbb{N}$. Each tube has section $s_k > 0$ and length $l_k > 0$. In the model, we shall neglect the geometrical constraints linked to the fact that tubes do not intersect. We shall denote the flow in the tube k by φ_k and the velocity of the fluid by v_k. One has $\varphi_k = s_k v_k$ and $\varphi_0 = 1$. The volume of the circuit is $V = \sum_k l_k s_k$ and we shall assume that this volume is prescribed and equal to V_0. We shall denote by (K) the set of constraints linking the φ_k's due the Kirchhoff's law. They read $\sum_{k \in I_j^+} \varphi_k = \sum_{k \in I_j^-} \varphi_k$ where j denotes an arbitrary vertex of the network and $I_j^\pm$ are respectively the sets of tubes arriving at and leaving j. It is not necessary to state explicitly the irrigation constraints (I) which ensure that the network irrigates a prescribed measure from a source S. In fact all calculations below can be made with a finite atomic irrigated measure. The network is entirely described by the finite list of its tubes defined by their axes. The axes build up a finite graph G embedded in $\mathbb{R}^3$. Thus the whole network along with its flow is defined by (G, l_k, s_k, φ_k) and in fact the l_k are a datum included in G. Assuming that the resistance R_k of a tube is an inverse power law of its section, $R_k = s_k^{-\beta}$, the energy of the network is defined by

$$W = \sum_k l_k s_k^{-\beta} s_k^2 v_k^2 = \sum_k l_k s_k^{-\beta} \varphi_k^2. \tag{14.2}$$

This energy can be interpreted as a dissipated power of the flow. The aim is to minimize this quantity, which leads to solve the variational problem with unknowns G, $s = (s_k)_k$, $\varphi = (\varphi_k)_k$,

$$\min_{(K),\ (I),\ V=V_0} (W(G, s, \varphi) = \sum_k l_k s_k^{-\beta} \varphi_k^2). \tag{14.3}$$

The volume constraint $V = V_0$ signals how much volume can be spent for the network. The aim of the next proposition is to prove by elimination of the unknown s_k and of the volume constraint $V = V_0$ that Problem (14.3) is in fact a Gilbert-Steiner problem with exponent $\alpha = \frac{2}{\beta+1}$.

Proposition 14.1. *Consider two finite atomic irrigated and irrigating probability measures. The optimal irrigation problem under volume constraint (14.3) is equivalent to the Gilbert-Steiner problem:*

$$\min_{(K),\ (I)} (\tilde{W}(G, \varphi) = \sum_k l_k \varphi_k^{\frac{2}{\beta+1}}). \tag{14.4}$$

Solutions (G, φ_k, s_k) for (14.3) and $(\tilde{G}, \tilde{\varphi}_k)$ to (14.4), if they exist, can be obtained from each other by setting $G = \tilde{G}$, $\varphi_k = \tilde{\varphi}_k$ and

$$s_k := \left(\frac{V_0}{\tilde{W}} \tilde{\varphi}_k^2 \right)^{\frac{1}{\beta+3}} .$$

Before proceeding with the proof, we notice that the model equivalence thus obtained is not quite satisfactory: we are not *a priori* allowed to move freely the radii in an optimal embedded circuit, since we do not take into account the fact that the tubes should not intersect. By Proposition 9.1, the existence of an optimal solution is granted for Problem (14.4). We do not intend to prove the existence of discrete optimal circuits mades of tubes for Problem (14.4). It is enough to show that if such solutions exist, they boil down to solutions of a Gilbert-Steiner problem. However, this existence can probably be obtained by extending the arguments of the proof of Proposition 9.1 and using the existence of optimal solutions for Problem (14.4).

Proof. Consider a minimal network for (14.3) and let us perform some Lagrangian calculus by perturbing the s_k. Leaving the flow in each tube φ_k fixed, one can perturb s_k provided the volume V does not change and the velocity v_k changes by the relation $s_k v_k = \varphi_k$. Thus perturbing all s_k's yields a feasible solution to (14.3) under the constraint $V = V_0$. By the Lagrange multiplier theorem there exists $\lambda \in \mathbb{R}$ such that $\frac{\partial W}{\partial s_k} = -\lambda \frac{\partial V}{\partial s_k}$, which yields $\beta s_k^{-\beta-1} \varphi_k^2 = \lambda$ and therefore $s_k = \left(\frac{\beta \varphi_k^2}{\lambda} \right)^{\frac{1}{\beta+1}}$. Substituting this expression in W we obtain

$$W(G, s, \varphi) = \left(\frac{\lambda}{\beta} \right)^{\frac{\beta}{\beta+1}} \sum_k l_k \varphi_k^{\frac{2}{\beta+1}} . \tag{14.5}$$

On the other hand

$$V_0 = \sum_k l_k s_k = \left(\frac{\beta}{\lambda} \right)^{\frac{1}{\beta+1}} \sum_k l_k \varphi_k^{\frac{2}{\beta+1}} = \frac{\beta}{\lambda} W.$$

Thus, we obtain the value of the Lagrange multiplier in terms of the energy and volume

$$\lambda = \frac{\beta W}{V_0}, \tag{14.6}$$

and therefore

$$s_k = \left(\frac{V_0}{W} \varphi_k^2 \right)^{\frac{1}{\beta+1}} . \tag{14.7}$$

These expressions of s_k and λ permit to get W in terms of V_0 and the fluxes φ_k only. Indeed, using (14.6) and (14.5),

$$W = \left(\frac{W}{V_0}\right)^{\frac{\beta}{\beta+1}} \sum_k l_k \varphi_k^{\frac{2}{\beta+1}},$$

which yields

$$W = V_0^{-\beta} \left(\sum_k l_k \varphi_k^{\frac{2}{\beta+1}}\right)^{\beta+1}. \tag{14.8}$$

Taking into account that the constraints (K) and (I) are given in terms of the φ_k only, we are led to consider the problem

$$\min_{(K),\,(I)} \left(\tilde{W} = V_0^{-\beta} \left(\sum_k l_k \varphi_k^{\frac{2}{\beta+1}}\right)^{\beta+1}\right). \tag{14.9}$$

This problem is expressed in terms of G and its fluxes φ_k only because the constraints (K) and (I) only deal with G and the φ_k, not with the sections s_k. Observe that, by (14.8) we have

$$\min_{(K),\,(I)} \tilde{W} \le \min_{(K),\,(I),\,V=V_0} W. \tag{14.10}$$

Let $(\tilde{G}, \tilde{\varphi}_k, \tilde{l}_k)$ be a solution of (14.9) and $\tilde{W}$ its energy. Following (14.7) we can define a section for the tubes of G by

$$\tilde{s}_k := \left(\frac{V_0}{\tilde{W}} \tilde{\varphi}_k^2\right)^{\frac{1}{\beta+1}}.$$

Then the volume of the network is

$$\tilde{V} = \sum_k \tilde{l}_k \left(\frac{V_0}{\tilde{W}} \tilde{\varphi}_k^2\right)^{\frac{1}{\beta+1}} = V_0.$$

Thus the solution $(\tilde{G}, \tilde{\varphi}_k, \tilde{s}_k)$ satisfies all constraints of Problem (14.3). Now, by (14.8) we have

$$\min_{(K),\,(I)} \tilde{W} \le \min_{(K),\,(I),\,V=V_0} W.$$

So this inequality is an equality and Problems (14.3) and (14.9) are equivalent. Of course solving (14.9) is equivalent to solving a Gilbert problem:

$$\min_{(K),\,(I)} \sum_k l_k \varphi_k^{\frac{2}{\beta+1}},$$

which ends the proof.

14.3 A Consequence of the Space-filling Condition

When saying that χ is a pattern irrigating a measurable set of positive measure in $\mathbb{R}^N$ (from the source S), we understand that χ irrigates the measure $\mu^- = \mathbb{1}_E \lambda_N$ from $\mu^+ = |E| \delta_S$ (λ_N denotes the Lebesgue measure on $\mathbb{R}^N$.)

Let $f : [0, \infty) \to [0, \infty)$ be an increasing continuous function such that $f(0) = 0$.

Definition 14.2. *Let E be a measurable set of positive measure in $\mathbb{R}^N$, $S \in \mathbb{R}^N \setminus E$ and χ a pattern irrigating E from the source S. Let us parameterize each fiber $\chi(\omega)$ so that $\chi(\omega, 0) = x$, $\chi(\omega, T(\omega)) = S$. We say that χ is an embedded pattern if for every fiber ω there is an accessibility profile, namely a function $f(\omega, s)$ such that $f(\omega, 0) = 0$, $f(\omega, s) > 0$ for $s \in (0, T(\omega)]$ and*

$$\forall s \in [0, T(\omega)], B(\chi(\omega, s), f(\omega, s)) = \chi(\omega, s) + B(0, f(\omega, s)) \subseteq \mathbb{R}^N \setminus E. \quad (14.11)$$

If $E \subseteq \mathbb{R}^N$ is Lebesgue-measurable and $x \in \mathbb{R}^N$, the density of E at x is defined by

$$d(E, x) := \lim_{\rho \to 0+} \frac{|E \cap B(x, \rho)|}{|B(x, \rho)|},$$

provided this limit exists. By Lebesgue density theorem [64], $d(E, x) = 1$ at almost every point of E. We also set

$$\overline{d}(E, x) := \lim_{\rho \to 0+} \sup \frac{|E \cap B(x, \rho)|}{|B(x, \rho)|},$$

Proposition 14.3. *[27] Let E be a measurable set of positive measure in $\mathbb{R}^N$ and χ an embedded pattern irrigating E. Let $x = \chi(\omega, 0)$ be a Lebesgue point of E. Then $\limsup_{s \to 0+} f(\omega, s)/s = 0$. As a consequence*

$$\int_0^{T(\omega)} \frac{1}{f(\omega, s)} dr = \infty. \quad (14.12)$$

Proof. Since $\chi(\omega, s)$ is parametrized by its arc length, we have $\chi(\omega, \frac{r}{2}) \in \overline{B(x, \frac{r}{2})}$. As a consequence, $B(\chi(\omega, \frac{r}{2}), \frac{r}{2}) \subset B(x, r)$. For simplicity let us write $\chi(s)$ for $\chi(\omega, s)$ and $f(s)$ for $f(\omega, s)$.

If $f(\frac{r}{2}) < \frac{r}{2}$, then $B(\chi(\frac{r}{2}), f(\frac{r}{2})) \subset B(x, r)$, so that

$$B(\chi(\frac{r}{2}), f(\frac{r}{2})) \cap B(x, r) = B(\chi(\frac{r}{2}), f(\frac{r}{2})).$$

If $f(\frac{r}{2}) \geq \frac{r}{2}$, then

$$B(\chi(\frac{r}{2}), \frac{r}{2}) \subset B(\chi(\frac{r}{2}), f(\frac{r}{2})) \cap B(x, r).$$

In both cases, we have $B(\chi(\frac{r}{2}), f(\frac{r}{2})) \subset \mathbb{R}^N \setminus E$, hence

$$\frac{|(\mathbb{R}^d \setminus E) \cap B(x, r)|}{|B(x, r)|} \geq \frac{|B(\chi(\frac{r}{2}), f(\frac{r}{2})) \cap B(x, r)|}{|B(x, r)|} \geq \frac{\min(r/2, f(\frac{r}{2}))^N}{r^N}$$

Taking the limsup, the inequality yields

$$\overline{d}(\mathbb{R}^d \setminus E, x) \geq \frac{1}{2^N} \min(\limsup_{r \to 0+} f(r)/r, 1)^N.$$

Thus, $d(E, x) = 1$ implies that $\limsup_{r \to 0+} f(r)/r = 0$. Finally observe that for some $R > 0$, $\frac{f(r)}{r} < 1$ for all $r < R$; otherwise we would have $\limsup_{r \to 0+} f(r)/r \geq 1$. It follows that $\frac{1}{r} < \frac{1}{f(r)}$ for all $r < R$, and thus $\int_0^R \frac{1}{f(r)} dr = \infty$.

So we obtained a generic constraint on accessibility profiles, not taken into account in finite models, but handy in infinitesimal ones.

14.4 Source to Volume Transfer Energy

Let χ be an embedded pattern irrigating a bounded subset $E \subset \mathbb{R}^3$ of positive measure. Let us define the volume and the Poiseuille dissipated power of the embedded pattern.

$$V(\chi) := \int_\Omega \int_0^{T(\omega)} \frac{f(\omega, s)^2}{|\chi(\omega, s)|_\chi} d\omega ds;$$

$$W^\beta(\chi) := \int_E \int_0^{T(\omega)} |\chi(\omega, s)|_\chi f(\omega, s)^{-2\beta}.$$

These expressions are direct extensions of the expressions of volume and energy for a single tube with constant section under Poiseuille's law (see Appendix B). They must be understood as lower bounds for the real V and W and it is easily checked that for all tube irrigating models, profiles $f(\omega)$ can be found so that these lower bounds are correct. The Poiseuille's law in dimension 3 corresponds to $\beta = 2$.

Theorem 14.4. *Let $\beta \geq 2$. For every embedded pattern irrigating a set $E \subset \mathbb{R}^3$ with positive measure, the energy W and the volume V of the network cannot be simultaneously finite.*

Proof. Without loss of generality, we assume that $|E| = 1$. Since $\mu = \pi_{\infty \sharp} \lambda$ (where λ is the Lebesgue measure on Ω), by Proposition 14.3, we have that

$$\int_0^{T(\omega)} \frac{1}{f(\omega, s)} dr = \infty \tag{14.13}$$

holds for almost every ω. Thus, using Cauchy-Schwartz inequality and the expressions of V and W, one has

$$\sqrt{V}\sqrt{W} \geq \int_\Omega \int_0^{T(\omega)} f^{1-\beta}(\omega, s) = +\infty \text{ if } \beta \geq 2.$$

14.5 Final Remarks

The architecture of optimal transport networks is the object of intensive study in many different contexts including water or gas networks, or the bronchial or vascular networks in animal physiology [34], [57]. The structure of networks is derived from an optimality principle, networks minimize the global resistance to flow (or dissipative energy) or the biological work with respect to different constraints (fixed total channel volume or total channel surface area). Of particular interest are the branching geometry, the relations between diameters and angles in bifurcations or the relation between the cross-sectional areas of adjoining channels at each junction (see [34] where this is studied for an optimal network with fixed topology). The node connectivity in bidimensional minimal resistance networks has been studied in [34] (see also [54]) where the author shows that no more than three or four channels meet at each junction, depending on the flow profile (e.g., Poiseuille-like or pluglike) and the considered constraint (fixed volume or surface area). The study of the arquitecture of the vascular network was initiated in [66], [65] based on the principle that, subject through evolution to natural selection, it must have achieved an optimum arrangement corresponding to the least possible biological work needed for maintaining the blood flow through it at required level. Again the branching properties are object of interest [57], [92]. The study of the optimal dimensioning of fluid (water or gas) networks when the topology is known leads to many interesting optimization problems in which the control of the pressure to the final user is required [30], [68], [54]. In [54] this leads to a control problem where the pressure and flow are constrained by Saint-Venant's equations. Finally, let us mention that the local optimization with respect to the network topology requires the use of topological derivative techniques [55].

15 Open Problems

15.1 Stability

Problem 15.1. Let $\chi_n \in TP(\mu_n^+, \mu_n^-)$ be a sequence of optimal traffic plans such that $E^\alpha(\chi_n)$ is uniformly bounded and χ_n converges to χ. Is χ optimal?

This is proved in Section 6.3 for $\alpha > 1 - \frac{1}{N}$.

See also the work [14] where the stability is considered in terms of variations of the parameter α rather than in terms of the measures μ^+ and μ^-.

Problem 15.2. Let μ^+ and μ^- be finite atomic measures and χ an optimal traffic plan from μ^+ to μ^- with finite graph structure of topology $\mathcal{T}$. Is it true that generically, for locally perturbed $\tilde{\mu}^+$, $\tilde{\mu}^-$, there is an optimal traffic plan from $\tilde{\mu}^+$ to $\tilde{\mu}^-$ that has the topology $\mathcal{T}$ (here, generically means that this property would be true for almost all finite atomic measures μ^+ and μ^-)?

This can be proved in dimension 2 with the Gilbert Steiner construction of local optima.

15.2 Regularity

Problem 15.3. Let μ^+ and μ^- be probability measures and $\chi \in TP(\mu^+, \mu^-)$ an optimal traffic plan. Does the interior regularity hold? That is to say, is it true that in any ball $B(x, r)$ outside of the support of the measures μ^+ and μ^-, the traffic plan χ has a finite graph structure?

This result has been proved for general $\alpha > 1 - \frac{1}{N}$ in the case μ^+ is a finite atomic measure (see Corollary 8.17). It has also been proved for $\alpha > 1 - \frac{1}{N}$, if μ^+ and μ^- have disjoint supports (see Theorem 8.14). It is then natural to ask whether it is possible to avoid both the conditions on α and on the supports of the measures.

Problem 15.4. Let $\chi \in TP(\mu^+, \mu^-)$ be an optimal traffic plan. Does χ have a tangent cone at any point of its support?

Let us mention that Xia proves the existence of a tangent cone for some subsequence.

M. Bernot et al., *Optimal Transportation Networks*. Lecture Notes in Mathematics 1955.
© Springer-Verlag Berlin Heidelberg 2009

15.3 The who goes where Problem

Problem 15.5. Let $\chi \in TP(\pi)$ be an optimal traffic plan for the "who goes where" problem. Is it true that in any ball $B(x, r)$ outside of the support of the measures μ^+ and μ^-, the traffic plan χ has a finite graph structure?

The only regularity available for now, is the one of the finite graph structure of χ if π has finite atomic marginals.

Problem 15.6. Let $\chi \in TP(\pi)$ be an optimal traffic plan for the "who goes where" problem. Is the number of branches at a branching point uniformly bounded?

Let us mention that the bi-Lipschitz regularity is true.

Problem 15.7. Let $\chi \in TP(\pi)$ be an optimal traffic plan for the "who goes where" problem. Is it approximable by finite graphs?

Problem 15.8. Let $\chi \in TP(\pi)$ be an optimal traffic plan for the "who goes where" problem, where π has finite atomic marginals. Find an algorithmic construction of the graph.

Problem 15.9. Find another multiplicity/energy model for the "who goes" problem.

In the current formulation, two paths going in opposite directions are allowed. For instance, people from town A travel to town B and people from town B travel simultaneously to town A. Yet, the current cost functional is such that the multiplicity is the same no matter whether the paths are going the same way or opposite ways. It is not easy to generalize the current study to the case where one defines a multiplicity that discriminates different directions at a point. In that case one could have different costs for two way roads and one way roads which have the same amount of traffic.

Another approach taken by Bernot and Figalli in [14] permits to distinguish the two multiplicities in the two towns example. To do so, they interpret the fiber $\chi(\omega, \cdot)$ as the trajectory of the particle ω and introduce a multiplicity that counts the proportion of particles ω going through a point at a given time. Thus, this multiplicity does not discriminate fiber directions but is time dependant. The energy based on this multiplicity is then related to the energy studied in this book and optimal traffic plans for these two costs are proved to be the same (up to a synchronization reparameterization), in the case μ^+ is atomic.

15.4 Dirac to Lebesgue Segment

Problem 15.10. Let χ be an optimal traffic plan from a Dirac mass to the Lebesgue measure on the segment. Consider χ_0 the connected components of χ outside the segment. We denote by μ_0 the measure irrigated by χ_0. What is the structure of μ_0? In particular, is it the Lebesgue measure on the segment, does μ_0 have some Dirac masses or is it made completely of Dirac masses? Does it have a Cantorian part?

See Chapter 13 for more details. In particular, Corollary 13.11 states that μ_0 can't be finite atomic for $\alpha > 0$.

15.5 Algorithm or Construction of Local Optima

Problem 15.11. Let μ^+ and μ^- be finite atomic measures in $X \subset \mathbb{R}^2$, and χ an optimal traffic plan (that has graph like structure because of regularity results) from μ^+ to μ^-. Find a construction algorithm of the optimal graph.

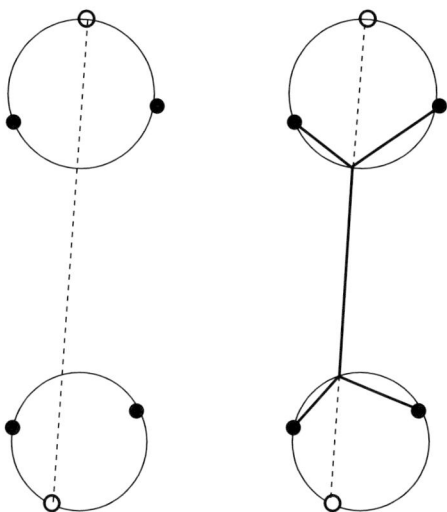

Fig. 15.1. The pivot point approach can give the optimum in that case since the top-dow, bottom-up approach holds in that case. Indeed, the construction of pivot points bring back the problem to a one source one target problem. We then reconstruct the whole structure as described in section 12.2.

Let us illustrate the difficulties appearing in the case of several sources. In case the optimal structure has a point S with multiplicity 1, the structure is the union of an optimal irrigation from S to μ^+ and an optimal irrigation from S to μ^- so that the pivot point approach holds (see Figure 15.1). However, if we try to find the optimal structure with the prescribed topology like the one represented on Figure 15.2, then the pivot point algorithm is of no use. Indeed, as illustrated by Figure 15.2, Steiner points are no longer obtained from top to down. The point b_1 depends on the location of b_2 and the point

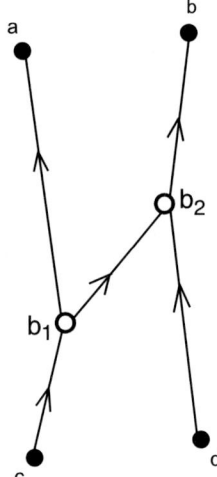

Fig. 15.2. If this structure is optimal, then the pivot points are of no help in finding the location of the bifurcation points b_1 and b_2. Indeed, we need b_1 to locate b_2 and reciprocally so that a numerical algorithm seems necessary in that case.

b_2 depends on the location of b_1. This calls for another approach and another coding of topologies.

Problem 15.12. Let μ^+ and μ^- be finite atomic measures in $X \subset \mathbb{R}^N$, with $N > 2$ and χ an optimal traffic plan (that has graph like structure because of regularity results) from μ^+ to μ^-. Find a construction algorithm of the optimal graph.

One main difficulty is added in the case of three dimensions: it is no more possible to use a combinatorial approach, even to optimize the transportation of a Dirac mass to a measure with very few Dirac masses. Let us go back to dimension 2 to explain that. In the case of 2 dimensions, each couple of points (P_1, P_2) can be reduced either to one of the two possible pivot points, either to P_1 or P_2. An exhaustive search through all possible topologies and all possible degeneracies can then take place.

In the case of 3 dimensions, given two points (P_1, P_2) the set of possible location for the pivot point is a whole circle. Thus, even for a prescribed topology, the combinatorics is of no help and one has to use numerical approximation.

15.6 Structure

Problem 15.13. Let $\chi \in TP(\mu^+, \mu^-)$ be an optimal traffic plan. Is it true that the length of fibers $\chi(\omega, \cdot)$ is uniformly bounded?

Problem 15.14. Let $\chi \in TP(\mu^+, \mu^-)$ be an optimal traffic plan, where μ^+ and μ^- are finite atomic measures. Is it true that no more than three branches can meet at a branching point?

This has been proved in Section 12.3 in dimension 2 for $\alpha \leq \frac{1}{2}$.

Problem 15.15. Let $\chi \in TP(\mu^+, \mu^-)$ be an optimal traffic plan, where μ^+ is a Dirac mass and μ^- is the Lebesgue measure on a unit square. One calls basin the measure irrigated by a branch. Is it true that basins are closed sets whose boundary is a null set (one says that the basin is essentially open). Also, given two adjacent basins, study the regularity of their boundary.

15.7 Scaling Laws

Problem 15.16. Let $\chi \in TP(\mu^+, \mu^-)$ be an optimal traffic plan, where μ^+ is a Dirac mass and μ^- is the Lebesgue measure on a unit segment. Given a path γ from the source to a point on the segment we know that it is piecewise linear (because of the interior regularity) and that it has an infinite number of branching points. Let us define l_n the length of the path between two successive branching points. Does $\frac{1}{n} \lim_{n \to \infty} \log l_n$ have a limit? One can ask for the same kind of scaling law for successive flows.

Problem 15.17. Let $\chi \in TP(\mu^+, \mu^-)$ be an optimal traffic plan, where μ^+ is a Dirac mass and μ^- is the Lebesgue measure on a unit square. Let x outside the support of χ. Is there a limit as $r \to 0$ for $\frac{1}{\log r} \log(w(r))$ where $w(r)$ is the sup or the average of the multiplicity on $S_\chi \cap B(x, r)$?

15.8 Local Optimality in the Case of Non Irrigability

Problem 15.18. Let μ^+ and μ^- be two probability measures and α_c the critical α value for the irrigation problem from μ^+ to μ^-. We say that χ is locally optimal if any pruning of χ is optimal. Prove that there always exist locally optimal traffic plans for $\alpha \in [0, \alpha_c]$.

Problem 15.19. Let μ^+ and μ^- be two probability measures and α_c the critical α value for the irrigation problem from μ^+ to μ^-. Let $\chi \in TP(\mu^+, \mu^-)$ be a locally optimal traffic plan for α_c. Is it true that χ is the limit of optimal traffic plans $\chi_n \in TP(\mu^+, \mu^-)$ for $\alpha_n \geq \alpha_c$ converging to α_c?

A Skorokhod Theorem

In this whole appendix (K, d) is a precompact metric space equipped with the σ–algebra of its Borel sets. Our aim here is to prove two theorems due to Skorokhod. The first one asserts that we can parameterize any borelian probability measure μ on K as the push-forward $\mu = \chi_\sharp \lambda$ of the Lebesgue measure λ on $[0, 1]$ by some borelian application $\chi : [0, 1] \to K$. The same construction can be extended to prove that the weak convergence of a sequence of probability measures μ_n to a probability measure μ on K is equivalent to the pointwise convergence of suitable parametrizations χ_n of μ_n towards a parameterization χ of μ.

Definition A.1. *Let (K, d) be a precompact metric space and μ be a probability measure on K. We call parameterization of μ a measurable application $\chi : \omega \in [0, 1] \to K$ such that $\mu = \chi_\sharp \lambda$ where λ is the Lebesgue measure on $[0, 1]$. That is to say $\mu(A) = \lambda(\chi^{-1}(A))$.*

Remark A.2. As a direct consequence of a parameterization χ of μ, if $\phi : K \to \mathbb{R}^+$ is a μ–measurable function, then $\int_K \phi(\gamma) d\mu(\gamma) = \int_\Omega \phi(\chi(\omega)) d\omega$ (see [5], Def. 1.70, p. 32). As an illustrative example the Dirac mass at 0 is parameterized by the null constant application on $[0, 1]$. In the same way, an atomic measure $\sum_1^n a_i \delta_{x_i}$ can be parameterized by the piecewise constant function $\chi(\omega) = x_1$ on $[0, a_1]$, $\chi(\omega) = x_2$ on $]a_1, a_2]$ and so on.

Theorem A.3. *(Skorokhod Theorem) Let (K, d) be a precompact metric space and μ be a probability measure on K. Then there exists a parameterization χ of μ.*

Theorem A.3 follows directly from Lemma A.6 where a more specific construction is achieved.

Lemma A.4. *There exists a filtration of K made of finite partitions $\mathcal{F}_l = \{F_j^l : 1 \leq j \leq J_l\}$, where $J_l \in \mathbb{N}^*$, such that the diameters of the sets F_j^l are less than 2^{-l}.*

Proof. We construct this filtration recursively. In order to construct $\mathcal{F}_1$, we cover K with a finite number of balls of radii $1/4$. Let us denote by B_i, where $1 \leq i \leq n$, the intersection of these balls with K. Let us find a partition of

$K = \cup_i B_i$ with at most n elements. To do this, we denote $\tilde{F}_1^1 := B_1$ and, in a recursive way, we define $\tilde{F}_{i+1}^1 := B_{i+1} \setminus \cup_{j \leq i} B_j$. If any of the $\tilde{F}_i^1$ is empty, we do not take it into account, so that we obtain a family of non empty elements F_i^1 where $i \leq J_1$. Since the F_i^1 are precompact, we can iterate the above process by covering them with balls of radius $1/8$. Proceeding iteratively we construct the desired filtration.

Lemma A.5. *Let μ be a probability measure on K. There exists a filtration made of finite partitions $\mathcal{F}_l = \{F_j^l : 1 \leq j \leq J_l\}$, $J_l \in \mathbb{N}^*$, such that the diameters of F_j^l are less than 2^{-l+1} and $\mu(\partial F_j^l) = 0$ for all l and $j \leq J_l$.*

Proof. To obtain this filtration, we slightly modify the construction of Lemma A.4. We only need to request in addition that $\mu(\partial F_j^l) = 0$ for all l and $j \in J_l$. For that, it is enough to perturb the radii $r_l = 2^{-l}(1 + \epsilon_l)$, with $\epsilon_l \leq 1$ so that μ does not charge the boundaries of the balls with radius r_l used to construct $\mathcal{F}_l$.

The filtration obtained in Lemma A.5 allows us to define a canonical parameterization of μ. The idea is to group together the ω's whose images are close.

Lemma A.6. *Let μ be a probability measure on K and $\mathcal{F}$ be the filtration constructed in Lemma A.5. There exists a parameterization χ of μ such that for all l, the sets*

$$\Omega_{j,l} = \{\omega : \chi(\omega) \in F_j^l\}$$

are intervals ordered in an increasing way with j.

Proof. We construct χ by successive approximations χ_n using the filtration of Lemma A.5.

Step 1: Definition of χ_n. Let $t_0^n := 0$ and $t_j^n := \sum_{i \leq j} \mu(F_i^n)$ where $1 \leq j \leq J_n$. The application χ_n is defined as a piecewise constant function sending each interval $[t_{j-1}^n, t_j^n[$ onto an arbitrary element of F_j^n. By construction, $\Omega_{j,l} := \{\omega : \chi_n(\omega) \in F_j^l\} = [t_{j-1}^l, t_j^l[$ for all $j \leq J_l$. We notice that the intervals $[t_{j-1}^l, t_j^l[$ where $1 \leq j \leq J_l$, are intervals ordered in an increasing way when j goes from 1 to J_l, so that its union is $[0, 1[$. Notice also that $\mu(F_j^l) = |\Omega_{j,l}|$.

Step 2: The sequence $\chi_n(\omega)$ converges for all ω. Let us prove that χ_n is a Cauchy sequence. Let us first observe that $\chi_n(\Omega_j^m) \subset F_j^m$ for any $n \geq m$. Indeed, let us fix m and $n \geq m$. By the definition of filtration, Ω_j^m is the union of Ω_k^n where k describes the set of indices such that $F_k^n \subset F_j^m$. Thus, χ_n sends every element of Ω_k^n to an element of $F_k^n \subset F_j^m$. A fortiori, the image of Ω_j^m under χ_n is in F_j^m. Now, since the sets F_j^m have diameter less than 2^{-m}, we deduce that $d(\chi_n(\omega), \chi_m(\omega)) < 2^{-m}$ for all $m \leq n$. Thus, $\chi_n(\omega)$ is a Cauchy sequence.

Let χ be the pointwise limit of χ_n. Observe that χ is measurable as a pointwise limit of measurable functions.

Step 3: The measure $\chi_\sharp \lambda$ is exactly μ. We have to show that $\chi_\sharp \lambda(F_j^l) = \mu(F_j^l)$ for all (j, l). The measures μ and $\chi_\sharp \lambda$ will then be equal on the sets F_j^l which form a Π-system. Then the extension Theorem of Π-systems (Lemma 1.6, p.19 [93]) shows that $\mu = \chi_\sharp \lambda$ on the σ-algebra generated by this Π-system, that is, on the σ-algebra of Borel sets of K.

Let us fix l, $j \leq J_l$, and let us define

$$G_p := \{\gamma \in F_j^l : d(\gamma, \partial F_j^l) \geq 1/p\}.$$

This is a non decreasing sequence of sets such that $\cup_p G_p = F_j^l \setminus \partial F_j^l$. Fix $\epsilon > 0$. For a sufficiently large p, we have

$$\mu(G_p) \geq \mu(F_j^l) - \epsilon. \tag{A.1}$$

Now, consider an l' such that $2^{-l'} < \frac{1}{2p}$. For any $y \in G_p$, there exists k so that $y \in F_k^{l'}$. Since the diameter of $F_k^{l'}$ is less than $\frac{1}{2p}$, $F_k^{l'} \subset G_{2p}$ so that $\bar{F}_k^{l'} \subset F_j^l$. For $n \geq l'$, the construction of χ_n ensures that $\chi_n(\Omega_k^{l'}) \subset F_k^{l'}$. Since χ is the pointwise limit of χ_n,

$$\chi(\Omega_k^{l'}) \subset \bar{F}_k^{l'} \subset F_j^l. \tag{A.2}$$

We obtain a covering of G_p with sets of the form $F_k^{l'}$ satisfying (A.2), and, using (A.1), we have $\chi_\sharp \lambda(F_j^l) \geq \mu(F_j^l) - \epsilon$. This being true for all $\epsilon > 0$, we deduce that $\chi_\sharp \lambda(F_j^l) \geq \mu(F_j^l)$. Since these sets form a partition for $1 \leq j \leq J_l$, and $\chi_\sharp \lambda$ is a probability measure, the inequality is indeed an equality, that is : $\chi_\sharp \lambda(F_j^l) = \mu(F_j^l)$. As a consequence, we have $\chi^{-1}(F_j^l) = \Omega_{j,l}$ modulo a null set.

Definition A.7. *Let $(\mu_n)_n$ and μ be probability measures on (K, d). We say that μ_n tends to μ "pointwise" whenever there exist parameterizations χ_n and χ of μ_n and of μ, respectively, such that $d(\chi_n(\omega), \chi(\omega)) \to 0$ almost everywhere in $[0, 1]$.*

Theorem A.8 (Skorokhod convergence Theorem). *Let $(\mu_n)_n$ be a sequence of probability measures on (K, d). The sequence μ_n weakly-$*$ converges to μ if and only if μ_n to μ tends to μ "pointwise".*

Proof. Assume that μ_n converges to μ "pointwise", and let χ_n, χ denote the parameterizations of μ_n and μ, respectively. Since $\chi_n(\omega)$ converges to $\chi(\omega)$ for almost every ω, using Lebesgue's theorem, for all $\phi \in C(K)$, we have

$$< \mu_n, \phi > = \int_K \phi(\gamma) d\mu_n(\gamma) = \int_{[0,1]} \phi(\chi_n(\omega)) d\omega$$

$$\to \int_{[0,1]} \phi(\chi(\omega)) d\omega = \int_K \phi(\gamma) d\mu(\gamma) = < \mu, \phi > .$$

Conversely, let μ_n be weakly-* converging to μ. Let us consider the filtration associated with μ constructed in Lemma A.5. Since $\mu(\partial F_j^l) = 0$, we deduce that $\mu_n(F_j^l)$ converges to $\mu(F_j^l)$. Next, applying Lemma A.6 to measures μ_n and μ, we get applications χ_n and χ such that $\chi_{n\sharp}\lambda = \mu_n$ and $\chi_\sharp\lambda = \mu$. The fact that $\mu_n(F_j^l)$ converges to $\mu(F_j^l)$ implies that $|\Omega_{j,l}^n|$ converges to $|\Omega_{j,l}|$, where $\Omega_{j,l}^n := \{\omega : \chi_n(\omega) \in F_j^l\}$ and $\Omega_{j,l} := \{\omega : \chi(\omega) \in F_j^l\}$. This convergence of measures implies the convergence of intervals $\Omega_{j,l}^n$ to some intervals $\Omega_{j,l}$, ordered in an increasing way with j.

We are now in a position to prove that for almost all ω the sequence $\chi_n(\omega)$ converges to $\chi(\omega)$. Notice that for almost all ω and for any $l \in \mathbb{N}$, there exists a $j \leq J_l$ such that ω is in the interior of $\Omega_{j,l}$. Indeed, there is a finite number of such intervals at each rank of the filtration, and thus, the set of its endpoints is countable, hence of measure zero. Thus, for n large enough, we have that $\omega \in \Omega_{j,l}^n$, i.e., $\chi_n(\omega) \in F_j^l$. This yields $d(\chi_n(\omega), \chi(\omega)) < 2^{-l}$.

B Flows in Tubes

In this appendix, we shall consider a fluid with laminar flow in a tube. We recall how Poiseuille's law can be derived from Navier-Stokes equation. Next, we discuss the optimality of the circular section.

B.1 Poiseuille's Law

Let us consider a tube of constant circular section with a straight axis. We take (x, y, l) as coordinates in the tube, where $l \in [0, L]$ is the distance along the axis and $(x, y) \in D(0, r)$ are orthogonal cartesian coordinates.

We assume a stationary regime and that the flow is laminar, that is to say the velocity is oriented by the axis and is constant on all trajectories, so that $\frac{\partial p}{\partial x} = \frac{\partial p}{\partial y} = 0$. The velocity v at a point of a tube along the z-axis is given by Navier-Stokes equation

$$- \triangle v(l)(x, y) - \frac{1}{\eta} \frac{\partial p}{\partial z}, \quad \text{where } \triangle = \frac{\partial^2}{\partial x^2} + \frac{\partial^2}{\partial y^2}$$

Hence, $\frac{\partial p}{\partial z} = constant$ (where η denotes the viscosity coefficient). Thus, the gradient of pressure has the form $\frac{[p]}{L}$ where $[p]$ denotes the pressure difference at the ends of the tube, and we shall denote it by $\bigtriangledown p$. In other words, p is a linear interpolation of the initial and final pressures in the tube. We assume that the pressure is constant on the initial and ending sections of the tube, so that the pressure is constant on each section of the tube. For simplicity, let us take $\eta = 1$.

Under these hypotheses, we can calculate the velocity and the corresponding flow through the whole tube

$$v(x, y, l) = \frac{(r^2 - (x^2 + y^2))}{4} \bigtriangledown p$$

$$f = \int_{D(0,r)} v(x, y, l) = \frac{1}{4} r^4 \bigtriangledown p = r^2 v_{max}$$

The power dissipated by the steady flow is $W = fL \bigtriangledown p$. This is to be identified with $W = Lf^2 R$ where by definition R stands for the resistivity of the tube. Thus we obtain $R = 4/r^4$: Poiseuille's law says that the resistivity of a tube scales as the inverse fourth power of the radius.

B.2 Optimality of the Circular Section

What is the optimal form of the section of a tube? If we prescribe the pressure at both ends of a tube of constant section, the circular form ensures the maximal flow. We briefly present the result obtained in [83] and [1].

Let us recall the definition of the rearrangement of a set (see [56]). If $A \subset \mathbb{R}^d$, we denote by A^* the ball $B(0, r) = \{x \mid |x| < r\}$ such that $|B(0, r)| = |A|$. If $f : \mathbb{R}^d \to \mathbb{R}$ is a Borel measurable function vanishing at infinity, we define the symmetric decreasing rearrangement of f by $f^*(x) = \int_0^\infty \chi^*_{\{|f|>t\}}(x) dt$. It results from the definition that $|\{x \mid |f(x)| > t\}| = |\{x \mid f^*(x) > t\}|$ and $||f||_p = ||f^*||_p$.

Let u be such that $-\triangle u(x, y) = \nabla p$ in the domain Ω. Let v be such that $-\triangle v(x, y) = (\nabla p)^* = \nabla p$ in Ω^*. Then, it can be shown that $u^* \leq v$ [83]. As a consequence, the flow in a tube of section Ω is such that $\int_\Omega u = \int_{\Omega^*} u^* \leq \int_{\Omega^*} v$. Then a circular section is always more advantageous from the point of view of the flow.

In [1], the authors prove the uniqueness of the optimal form : if $\max u = \max v$, then there is x_0 such that $\Omega = x_0 + \Omega^*$ and $u = v(\cdot + x_0)$. Then, if Ω is an optimal form, we have $\int_\Omega u = \int_{\Omega^*} v$ and $u^* \leq v$, hence $\max u = \max u^* = \max v$ necessarily. Then there is x_0 such that $\Omega = x_0 + \Omega^*$, and, therefore, the circular form is the unique optimum.

C Notations

$\mathcal{C}(X,Y)$ set of continuous maps from X to Y endowed with sup-norm

X convex compact set of $\mathbb{R}^N$

$\mathcal{P}(X)$ metric space of probabilities on X

K set of 1-Lipschitz curves or paths $\mathbb{R}^+ \to X$, page 25

$L(\gamma)$ length of path, page 31

$T(\gamma)$ stopping time of a path, page 25

$\mathcal{C}([a,b],\mathbb{R}^N)$ set of continuous curves from $[a,b]$ to $\mathbb{R}^N$, endowed with the sup-norm $\|\cdot\|$

$\mathcal{C}_b(\mathbb{R}^N,\mathbb{R}^N)$ set of continuous bounded functions from $\mathbb{R}^N$ to $\mathbb{R}^N$

δ_x, Dirac mass at x

$|A|$ Lebesgue measure of a measurable set $A \subset \mathbb{R}$

$\lambda(A)$ Lebesgue measure of a measurable set $A \subset \mathbb{R}$

$conv(E)$ the convex hull of E

$E \Delta F = (E \setminus F) \cup (F \setminus E)$ symmetric difference of sets

$\mathcal{H}^1$ one-dimensional Hausdorff measure (length)

$M^\alpha(G) = \sum_{e \in E(G)} f(e)^\alpha \mathcal{H}^1(e)$ Gilbert energy, page 13

$M^\alpha(\mu^+, \mu^-)$ minimal Gilbert-Xia transport energy from μ^+ to μ^-, page 13

$\tilde{E}^\alpha(\chi)$ energy of a pattern page, 14

$\Omega = [0,1]$ or a fixed subset of $\mathbb{R}$ with finite measure, set of fiber indices

$\omega \in \Omega$ a fiber index

$[\omega]_t$ branch of ω at time t, page 14

μ^+, μ^- positive Borel measures in X with equal mass

π a probability measure on $X \times X$ or "transference plan"

$\boldsymbol{P}$ traffic plan, page 26

$\pi_{\boldsymbol{P}}$ transference plan associated with a traffic plan $\boldsymbol{P}$, page 26

π_0, page 26

π_∞, page 26

$\pi_{0\sharp}\boldsymbol{P}$, $\pi_{\infty\sharp}\boldsymbol{P}$, page 26

$t \mapsto \chi(\omega, t)$, fiber indexed by ω

$(\omega, t) \mapsto \chi(\omega, t)$, parameterized traffic plan or pattern, pages 27, 28

$\boldsymbol{P}_\chi$ law in K of a parameterized traffic plan, page 28

$|\chi| := |\Omega|$ the total mass transported by χ

$T_\chi(\omega) := \inf\{t : \chi(\omega) \text{ is constant on } [t, \infty)\}$, page 28

$T(\omega)$ abbreviation for $T_\chi(\omega)$, page 28

$\tau(\omega) = \chi(\omega,0)$ initial point of a fiber page, 29
$\sigma(\omega) = \chi(\omega,T(\omega))$ final point of a fiber, page 29
$\pi = (\tau,\sigma)_\sharp\lambda$ transference plan of χ
$\mu^+(\chi)(A) := |\{\omega : \chi(\omega,0) \in A\}|$, irrigating measure of χ
$\mu^+_\chi = \tau_\sharp\lambda$, irrigating measure of χ
$\mu^-(\chi)(A) := |\{\omega : \chi(\omega,T_\chi(\omega)) \in A\}|$, measure irrigated by χ
$\mu^-_\chi = \sigma_\sharp\lambda$, measure irrigated by χ
$\mathrm{TP}(\mu^+,\mu^-)$ set of traffic plans χ such that $\mu^-_\chi = \mu^-$ and $\mu^+_\chi = \mu^+$, page 143
TP_C set of traffic plans such that $\int_\Omega T_\chi(\omega)\,d\omega \le C$, page 26
$\mathrm{TP}_C(\mu^+,\mu^-) := \mathrm{TP}(\mu^+,\mu^-) \cap \mathrm{TP}_C$.
$[\omega]_t \subset \Omega$ branch of the fiber ω at time t in a pattern, page 14
$|[\omega]_t|$ measure of the branch containing ω at time t, page 14
$\tilde{E}^\alpha(\chi) = \int_\Omega \int_0^{T(\omega)} |[\omega]_t|^{\alpha-1} d\omega dt$ energy of a pattern, page 14
$\mathcal{D}(\chi) = \cup_{\omega\in\Omega}\{\omega\} \times [S(\omega),T(\omega)]$ domain of a traffic plan, page 48
χ_D restriction of χ to a sub-domain of $\mathcal{D}(\chi)$, page 47
$\Omega^\chi_x := \{\omega : x \in \chi(\omega,\mathbb{R})\}$ path class of x in χ, page 31
$\Omega_x := \Omega^\chi_x$ abbreviation
$|x|_\chi = |\Omega^\chi_x|$ multiplicity of χ at x, page 31
$|x|_P$ multiplicity of P at x, page 31
S_χ, S_P support of P or of its parameterized traffic plan χ, set of points x such that $|x|_\chi > 0$, page 31
$E^\alpha(\chi) = \int_\Omega \int_{\mathbb{R}^+} |x(\omega,t)|^{\alpha-1}_\chi |\dot{\chi}(\omega,t)| dt d\omega$ energy of a traffic plan P parameterized by χ, page 35
$E^\alpha(\mu^+,\mu^-) := \min_{\mathrm{TP}(\mu^+,\mu^-)} E^\alpha(\chi)$ minimal energy between μ^+ and μ^-, page 55
$\Omega_{\overrightarrow{xy}}$ set of fibers passing through x and then through y, page 65
χ_{xy} traffic plan made of all pieces of fibers of χ joining x to y, page 65
$\Gamma^{xy} = S_{\chi_{xy}}$ support of χ_{xy}, page 65
$\Omega_{xy}:=\Omega_{\overrightarrow{xy}} \cup \Omega_{\overrightarrow{yx}}$, page 70
Γ_{xy} the only arc with positive multiplicity between x and y, page 70

References

1. A. ALVINO, P.L. LIONS, and G. TROMBETTI. A remark on comparison results via symmetrisation. *Proceeding of the Royal Society of Edimburgh*, 102A:37–48, 1986.

2. L. AMBROSIO. *Mathematical aspects of evolving interfaces*, volume 1812 of *CIME Series*, chapter Lecture Notes on Optimal Transport Problems, pages 1–52. Springer, Madeira, P. Colli and J.F. Rodrigues edition, 2003.

3. L. AMBROSIO. Transport equation and Cauchy problem for BV vector fields. *Inventiones Mathematicae*, 158(2):227–260, 2004.

4. L. AMBROSIO. Transport equation and Cauchy problem for non-smooth vector fields. *Lecture Notes of the CIME Summer school in Cetrary*, 2005.

5. L. AMBROSIO, N. FUSCO, and D. PALLARA. *Functions of bounded variations and free discontinuity problems.* Oxford Mathematical Monographs. The Clarendon Press, Oxford University Press, 2000.

6. J.R. BANAVAR, F. COLAIORI, A. FLAMMINI, A. MARITAN, and A. RINALDO. Scaling, optimality, and landscape evolution. *J. Stat. Physics*, 104(1):1–48, 2001.

7. V. BANGERT. Minimal Measures and Minimizing Closed Normal Onecurrents. *Geometric And Functional Analysis*, 9(3):413–427, 1999.

8. A. BEJAN and M.R. ERRERA. Deterministic tree networks for fluid flow: Geometry for minimal flow resistance between a volume and one point. *Fractals*, 5(4):685–695, 1997.

9. P. BERNARD and B. BUFFONI. Optimal mass transportation and Mather theory. *J. Eur. Math. Soc*, 9(1):85–121, 2007.

10. M. BERNOT. Irrigation and optimal transport. *PhD Thesis, École Normale Supérieure de Cachan. Available at* http://umpa.ens-lyon.fr/~bernot, 2005.

11. M. BERNOT, V. CASELLES, and J.-M. MOREL. Traffic plans. *Publicacions Matemàtiques*, 49(2):417–451, 2005.

12. M. BERNOT, V. CASELLES, and J.-M. MOREL. Are there infinite irrigation trees? *Journal of Mathematical Fluid Mechanics*, 8(3):311–332, 2006.

13. M. BERNOT, V. CASELLES, and J.-M. MOREL. The structure of branched transportation networks. *Calc. Var. and PDE*, 32(3):279–317, 2008.

14. M. BERNOT and A. FIGALLI. Synchronized traffic plans and stability of optima. *to appear in ESAIM: COCV*, 2008.

15. A.S. BESICOVITCH. On the definition and value of the area of a surface. *Quart. J. Math.*, 16:86–102, 1945.

16. S. BHASHKARAN and F. J. M. SALZBORN. Optimal design of gas pipeline networks. *J. Oper. Res. Society*, 30:1047–1060, 1979.

194   References

17. A. BRANCOLINI and G. BUTTAZZO. Optimal networks for mass transportation problems. *ESAIM: COCV*, 11:88–101, 2005.

18. A. BRANCOLINI, G. G. BUTTAZZO, and F. SANTAMBROGIO. Path functionals over Wasserstein spaces. *J. Eur. Math. Soc.*, 8(3):415–434, 2006.

19. Y. BRENIER. The Least Action Principle and the Related Concept of Generalized Flows for Incompressible Perfect Fluids. *Journal of the American Mathematical Society*, 2(2):225–255, 1989.

20. J.H. BROWN, G.B. WEST, and B.J. ENQUIST. *Scaling in biology*, chapter The origin of universal scaling laws in biology, pages 87–112. Oxford University Press, 2000.

21. G. BUTTAZZO. *Three Optimization Problems in Mass Transportation Theory*, volume 12 of *Advances in Mechanics and Mathematics*, chapter Nonsmooth Mechanics and Analysis, pages 13–23. Springer, 2006.

22. G. BUTTAZZO, A. PRATELLI, S. SOLIMINI, and E. STEPANOV. Mass transportation and urban planning problems. *Forthcoming*.

23. G. BUTTAZZO, A. PRATELLI, and E. STEPANOV. Optimal pricing policies for public transportation networks. *SIAM J. Opt.*, 16:826–853, 2006.

24. G. BUTTAZZO and E. STEPANOV. Optimal transportation networks as free Dirichlet regions for the Monge-Kantorovich problem. *Ann. Sc. Norm. Super. Pisa*, 5(4):631–678, 2003.

25. G. BUTTAZZO and E. STEPANOV. *Minimization problems for average distance functionals*, volume 14, pages 47–83. 2004.

26. S. CAMPANATO. Proprieta di Holderianita di alcune classi di funzioni. *Ann. Scuola Norm. Sup. Pisa*, 17:175–188, 1963.

27. V. CASELLES and J.-M. MOREL. *Irrigation*, pages 81–90. Progress in Nonlinear Differential Equations and their Applications, vol. 51, VARMET 2001, Trieste, June, 2001. Birkhäuser, F. Tomarelli and G. Dal Maso edition, 2002.

28. G. DEVILLANOVA. Singular structures in some variational problems. *PhD Thesis, École Normale Supérieure de Cachan*, 2005.

29. G. DEVILLANOVA and S. SOLIMINI. Forthcoming. 2006.

30. D. DE WOLF and Y. SMEERS. Optimal dimensioning of pipe networks with application to gas transmission networks. *Operations Research*, 44(4):596–608, 1996.

31. A.K. DEB. Least cost design of branched pipe network system. *Journal of Environmental Engineering Division ASCE*, 100(4):821–835, 1974.

32. P.S. DODDS, D.H. ROTHMAN, and J.S. WEITZ. Re-examination of the 3/4-law of metabolism. *Journal of Theoretical Biology*, 209:9–27, 2001.

33. R.M. DUDLEY. *Real Analysis and Probability*. Cambridge University Press, 2002.

34. M. DURAND. Architecture of optimal transport networks. *Physical Review E*, 73:016116, 2006.

35. G. EIGER, U. SHAMIR, and A. BEN-TAL. Optimal design of water distribution networks. *Water Resources Research*, 30(9):2637–2646, 1994.

36. I. EKELAND and R. R. TEMAM. *Analyse Convexe et Problèmes Variationnels*. Études Mathématiques. Dunod, Gauthier-Villars, 1974.

37. L.C. EVANS. Partial differential equations and Monge-Kantorovich mass transfer. *Current Developments in Mathematics*, pages 65–126, 1997.

38. L.C. EVANS and W. GANGBO. *Differential equations methods for the Monge-Kantorevich mass transfer problem*. American Mathematical Society Providence, RI, 1999.

39. K.J. FALCONER. *The geometry of fractal sets*. Cambridge University Press, 1985.
40. H. FEDERER. *Geometric Measure Theory*. Classics in Mathematics. Springer Verlag, 1969.
41. D.B.M.M. FONTES, E. HADJICONSTANTINOU, and N. CHRISTOFIDES. A Branch-and-Bound Algorithm for Concave Network Flow Problems. *Journal of Global Optimization*, 34(1):127–155, 2006.
42. W. GANGBO and R.J. McCANN. The geometry of optimal transportation. *Acta Mathematica*, 177(2):113–161, 1996.
43. M. GIAQUINTA. *Multiple integrals in calculus of variations and nonlinear elliptic systems*. Princeton University Press, 1983.
44. E.N. GILBERT. Minimum cost communication networks. *Bell System Tech. J.*, 46:2209–2227, 1967.
45. G.M. GUISEWITE and P.M. PARDALOS. Minimum concave-cost network flow problems: Applications, complexity, and algorithms. *Annals of Operations Research*, 25(1):75–99, 1990.
46. G.M. GUISEWITE and P.M. PARDALOS. Algorithms for the single-source uncapacitated minimum concave-cost network flow problem. *Journal of Global Optimization*, 1(3):245–265, 1991.
47. R.E. HORTON. Erosional development of streams and their drainage basins; hydrophysical approach to quantitative morphology. *Geol. Soc. Am. Bull.*, 56:275, 1945.
48. E. IJJASZ-VASQUEZ, R.L. BRAS, I. RODRIGUEZ-ITURBE, R. RIGON, and A. RINALDO. Are river basins optimal channel networks? *Advances in Water Resources*, 16:69–79, 1993.
49. D. JUNGNICKEL. *Graphs, Networks And Algorithms*. Springer, 2004.
50. L. KANTOROVICH. On the transfer of masses. *Dokl. Acad. Nauk. USSR*, 37:7–8, 1942.
51. T. LARSSON, A. MIGDALAS, and M. RONNQVIST. A Lagrangian heuristic for the capacitated concave minimum cost network flow problem. *European Journal of Operational Research*, 78:116–129, 1994.
52. D.H. LEE. Low cost drainage networks. *Networks*, 6:351–371, 1976.
53. R.P. LEJANO. Optimizing the layout and design of branched pipeline water distribution systems. *Irrigation and Drainage Systems*, 20(1):125–137, 2006.
54. G. LEUGERING. Structural optimization and control for partial differential equations on networks. *Preprint*, 2007.
55. G. LEUGERING and J SOKOLOWSKI. *Topological derivatives for elliptic problems on graphs*. Variational formulations in Mechanics: theory and applications. CIMNE, Barcelona, Spain, E. Tarocco, E.A. de Souza Neto and A. A. Novotny edition, 2006.
56. E.H. LIEB and M. LOSS. *Analysis*, volume 14 of *Graduate Studies in Mathematics*. American Mathematical Society, 1997.
57. I.A. LUBASHEVSKY and V. V. GAFIYCHUK. Analysis of the optimality principles responsible for vascular network architectonics. *Am. J. Physiol.*, 1:209–213, 1999.
58. F. MADDALENA and S. SOLIMINI. Transport distances and irrigation problems. *preprint*, 2007.
59. F. MADDALENA, S. SOLIMINI, and J.-M. MOREL. A variational model of irrigation patterns. *Interfaces and Free Boundaries*, 5(4):391–416, 2003.

60. P. MATTILA. *Geometry of Sets and Measures in Euclidean Spaces Fractals and rectificability*. Cambridge University Press, 1995.

61. B. MAUROY, M. FILOCHE, E. WEIBEL, and B. SAPOVAL. An optimal bronchial tree may be dangerous. *Nature*, 427:633–636, 2004.

62. G. MONGE. Mémoire sur la théorie des déblais et de remblais. *Histoire de l'Académie Royale des Sciences de Paris*, pages 666–704, 1781.

63. J.M. MOREL and F. SANTAMBROGIO. Comparison of distances between measures. *Applied Mathematics Letters*, 20(4):427–432, 2007.

64. F. MORGAN. *Geometric measure theory. A beginner's guide*. Academic Press, 1995.

65. C.D. MURRAY. The physiological principle of minimum work applied to the angle of branching arteries. *J. Gen. Physiology*, 9:835–841, 1926.

66. C.D. MURRAY. The physiological principle of minimum work I. The vascular system and the cost of blood volume. *Proc. natl. Acad. Sci. USA*, 12:207–214, 1926.

67. W.I. NEWMAN, D.L. TURCOTTE, and A.M. GABRIELOV. Fractal trees with side branching. *Fractals*, 5(4):603–614, 1997.

68. A.J. OSIADACZ and M GÓRECKI. *Optimization of pipe sizes for distribution gas network design*. Proceedings of the 27th PSIG Annual Meeting, Albuquerque, EUA October. 1995.

69. E. PAOLINI and E. STEPANOV. Connecting measures by means of branched transportation networks at finite cost. *Preprint*, 2006.

70. E. PAOLINI and E. STEPANOV. Optimal transportation networks as flat chains. *Interfaces and Free Boundaries*, 8:393–436, 2006.

71. A. PRATELLI. Equivalence between some definitions for the optimal mass transport problem and for the transport density on manifolds. *Ann. Mat. Pura Appl.*, 184(2):215–238, 2005.

72. R. RIGON, A. RINALDO, I. RODRIGUEZ-ITURBE, E. IJJASZ-VASQUEZ, and R.L. BRAS. Optimal channel networks: a framework for the study of river basin morphology. *Water Resources Research*, 29(6):1635–1646, 1993.

73. I. RODRIGUEZ-ITURBE and A. RINALDO. *Fractal river basins, chance and self-organization*. Cambridge University Press, 2001.

74. I. RODRIGUEZ-ITURBE, A. RINALDO, R. RIGON, R.L. BRAS, E. IJJASZ-VASQUEZ, and A. MARANI. Fractal structures as least energy dissipation patterns: the case of river networks. *Geophysical Research Letters*, 5:2854–2860, 1992.

75. F. SANTAMBROGIO. Optimal channel networks, landscape function and branched transport. *Interface and Free Boundaries*, 9:149–169, 2007.

76. F. SANTAMBROGIO. Transport and Concentration Problems with Interaction Effects. *J. Glob. Opt.*, 38(1):129–141, 2007.

77. H.D. SHERALI and E.P. SMITH. A Global Optimization Approach to a Water Distribution Network Design Problem. *Journal of Global Optimization*, 11(2):107–132, 1997.

78. S. SOLIMINI and G. DEVILLANOVA. On the dimension of an irrigable measure. *Preprint*, 2005.

79. S. SOLIMINI and G. DEVILLANOVA. Elementary properties of optimal irrigation patterns. *Calc. Var. and PDE*, 28(3):317–349, 2007.

80. E.O. STEPANOV. Optimization model of transport currents. *Journal of Mathematical Sciences*, 135(6):3457–3484, 2006.

81. E.O. STEPANOV. Partial Geometric Regularity of Some Optimal Connected Transportation Networks. *Journal of Mathematical Sciences*, 132(4):522–552, 2006.

82. A.N. STRAHLER. Quantitative analysis of watershed geomorphology. *Am. Geophys. Un. Trans.*, 38:913, 1957.

83. G. TALENTI. Elliptic equations and rearrangements. *Ann. Scuola Norm. Sup. Pisa*, 3:697–718, 1976.

84. E. TOKUNAGA. Consideration on the composition of drainage networks and their evolution. *Geogr. Rep., Tokyo Metrop. Univ*, 13:1–27, 1978.

85. H. TUY, S. GHANNADAN, A. MIGDALAS, and P. VÄRBRAND. The Minimum Concave Cost Network Flow Problem with fixed numbers of sources and nonlinear arc costs. *Journal of Global Optimization*, 6(2):135–151, 1995.

86. C. VILLANI. *Topics in optimal transportation*, volume 58 of *Graduate Studies in Mathematics*. American mathematical society, Providence, RI, 2003.

87. C. VILLANI. Saint-Flour Lecture Notes. *Optimal transport, old and new.* Available online via http://umpa.ens-lyon.fr/~cvillani, 2005.

88. G.B. WEST. The origin of universal scaling laws in biology. *Physica A*, 263:104–113, 1999.

89. G.B. WEST, J.H. BROWN, and B.J. ENQUIST. A general model for the origin of allometric scaling laws in biology. *Science*, 276(4):122–126, 1997.

90. G.B. WEST, J.H. BROWN, and B.J. ENQUIST. A general model for ontogenetic growth. *Nature*, 413:628–631, 2001.

91. B. WHITE. Rectifiability of flat chains. *Annals of Mathematics*, 150:165–184, 1999.

92. P. W. A. WILLEMS, K. S. HAN, and . HILLEN. Evaluation by solid vascular casts of arterial geometric optimisation and the inuence of ageing. *J. Anat.*, 196:161–171, 2000.

93. D. WILLIAMS. *Probability with martingales*. Cambridge University Press, 1991.

94. Q. XIA. Optimal paths related to transport problems. *Commun. Contemp. Math.*, 5:251–279, 2003.

95. Q. XIA. Boundary regularity of optimal transport paths. *Preprint*, 2004.

96. Q. XIA. Interior regularity of optimal transport paths. *Calculus of Variations and Partial Differential Equations*, 20(3):283–299, 2004.

97. Q. XIA. The formation of a tree leaf. *ESAIM: COCV*, 13(2):359–377, 2007.

98. G. XUE, T. LILLYS, and D. DOUGHERTY. Computing the minimum cost pipe network interconnecting one sink and many sources. *SIAM J. Optim.*, 10:22–42, 2000.

99. S. YAN, D. JUANG, C. CHEN, and W. LAI. Global and Local Search Algorithms for Concave Cost Transshipment Problems. *Journal of Global Optimization*, 33(1):123–156, 2005.

100. W.I. ZANGWILL. Minimum Concave Cost Flows in Certain Networks. *Management Science*, 14(7):429–450, 1968.

101. J.Z. ZHANG and D.T. ZHU. A bilevel programming method for pipe network optimization. *SIAM J. Optim.*, 6:838–857, 1996.

Index

Lecture Notes in Mathematics

For information about earlier volumes
please contact your bookseller or Springer
LNM Online archive: springerlink.com

Vol. 1817: E. Koelink, W. Van Assche (Eds.), Orthogonal Polynomials and Special Functions. Leuven 2002 (2003)
Vol. 1818: M. Bildhauer, Convex Variational Problems with Linear, nearly Linear and/or Anisotropic Growth Conditions (2003)
Vol. 1819: D. Masser, Yu. V. Nesterenko, H. P. Schlickewei, W. M. Schmidt, M. Waldschmidt, Diophantine Approximation. Cetraro, Italy 2000. Editors: F. Amoroso, U. Zannier (2003)
Vol. 1820: F. Hiai, H. Kosaki, Means of Hilbert Space Operators (2003)
Vol. 1821: S. Teufel, Adiabatic Perturbation Theory in Quantum Dynamics (2003)
Vol. 1822: S.-N. Chow, R. Conti, R. Johnson, J. Mallet-Paret, R. Nussbaum, Dynamical Systems. Cetraro, Italy 2000. Editors: J. W. Macki, P. Zecca (2003)
Vol. 1823: A. M. Anile, W. Allegretto, C. Ringhofer, Mathematical Problems in Semiconductor Physics. Cetraro, Italy 1998. Editor: A. M. Anile (2003)
Vol. 1824: J. A. Navarro González, J. B. Sancho de Salas, $\mathscr{C}^\infty$ – Differentiable Spaces (2003)
Vol. 1825: J. H. Bramble, A. Cohen, W. Dahmen, Multiscale Problems and Methods in Numerical Simulations, Martina Franca, Italy 2001. Editor: C. Canuto (2003)
Vol. 1826: K. Dohmen, Improved Bonferroni Inequalities via Abstract Tubes. Inequalities and Identities of Inclusion-Exclusion Type. VIII, 113 p, 2003.
Vol. 1827: K. M. Pilgrim, Combinations of Complex Dynamical Systems. IX, 118 p, 2003.
Vol. 1828: D. J. Green, Gröbner Bases and the Computation of Group Cohomology. XII, 138 p, 2003.
Vol. 1829: E. Altman, B. Gaujal, A. Hordijk, Discrete-Event Control of Stochastic Networks: Multimodularity and Regularity. XIV, 313 p, 2003.
Vol. 1830: M. I. Gil', Operator Functions and Localization of Spectra. XIV, 256 p, 2003.
Vol. 1831: A. Connes, J. Cuntz, E. Guentner, N. Higson, J. E. Kaminker, Noncommutative Geometry, Martina Franca, Italy 2002. Editors: S. Doplicher, L. Longo (2004)
Vol. 1832: J. Azéma, M. Émery, M. Ledoux, M. Yor (Eds.), Séminaire de Probabilités XXXVII (2003)
Vol. 1833: D.-Q. Jiang, M. Qian, M.-P. Qian, Mathematical Theory of Nonequilibrium Steady States. On the Frontier of Probability and Dynamical Systems. IX, 280 p, 2004.
Vol. 1834: Yo. Yomdin, G. Comte, Tame Geometry with Application in Smooth Analysis. VIII, 186 p, 2004.
Vol. 1835: O.T. Izhboldin, B. Kahn, N.A. Karpenko, A. Vishik, Geometric Methods in the Algebraic Theory of Quadratic Forms. Summer School, Lens, 2000. Editor: J.-P. Tignol (2004)
Vol. 1836: C. Năstăsescu, F. Van Oystaeyen, Methods of Graded Rings. XIII, 304 p, 2004.
Vol. 1837: S. Tavaré, O. Zeitouni, Lectures on Probability Theory and Statistics. Ecole d'Eté de Probabilités de Saint-Flour XXXI-2001. Editor: J. Picard (2004)
Vol. 1838: A.J. Ganesh, N.W. O'Connell, D.J. Wischik, Big Queues. XII, 254 p, 2004.
Vol. 1839: R. Gohm, Noncommutative Stationary Processes. VIII, 170 p, 2004.
Vol. 1840: B. Tsirelson, W. Werner, Lectures on Probability Theory and Statistics. Ecole d'Eté de Probabilités de Saint-Flour XXXII-2002. Editor: J. Picard (2004)
Vol. 1841: W. Reichel, Uniqueness Theorems for Variational Problems by the Method of Transformation Groups (2004)

Vol. 1842: T. Johnsen, A. L. Knutsen, K_3 Projective Models in Scrolls (2004)
Vol. 1843: B. Jefferies, Spectral Properties of Noncommuting Operators (2004)
Vol. 1844: K.F. Siburg, The Principle of Least Action in Geometry and Dynamics (2004)
Vol. 1845: Min Ho Lee, Mixed Automorphic Forms, Torus Bundles, and Jacobi Forms (2004)
Vol. 1846: H. Ammari, H. Kang, Reconstruction of Small Inhomogeneities from Boundary Measurements (2004)
Vol. 1847: T.R. Bielecki, T. Björk, M. Jeanblanc, M. Rutkowski, J.A. Scheinkman, W. Xiong, Paris-Princeton Lectures on Mathematical Finance 2003 (2004)
Vol. 1848: M. Abate, J. E. Fornaess, X. Huang, J. P. Rosay, A. Tumanov, Real Methods in Complex and CR Geometry, Martina Franca, Italy 2002. Editors: D. Zaitsev, G. Zampieri (2004)
Vol. 1849: Martin L. Brown, Heegner Modules and Elliptic Curves (2004)
Vol. 1850: V. D. Milman, G. Schechtman (Eds.), Geometric Aspects of Functional Analysis. Israel Seminar 2002-2003 (2004)
Vol. 1851: O. Catoni, Statistical Learning Theory and Stochastic Optimization (2004)
Vol. 1852: A.S. Kechris, B.D. Miller, Topics in Orbit Equivalence (2004)
Vol. 1853: Ch. Favre, M. Jonsson, The Valuative Tree (2004)
Vol. 1854: O. Saeki, Topology of Singular Fibers of Differential Maps (2004)
Vol. 1855: G. Da Prato, P.C. Kunstmann, I. Lasiecka, A. Lunardi, R. Schnaubelt, L. Weis, Functional Analytic Methods for Evolution Equations. Editors: M. Iannelli, R. Nagel, S. Piazzera (2004)
Vol. 1856: K. Back, T.R. Bielecki, C. Hipp, S. Peng, W. Schachermayer, Stochastic Methods in Finance, Bressanone/Brixen, Italy, 2003. Editors: M. Fritelli, W. Runggaldier (2004)
Vol. 1857: M. Émery, M. Ledoux, M. Yor (Eds.), Séminaire de Probabilités XXXVIII (2005)
Vol. 1858: A.S. Cherny, H.-J. Engelbert, Singular Stochastic Differential Equations (2005)
Vol. 1859: E. Letellier, Fourier Transforms of Invariant Functions on Finite Reductive Lie Algebras (2005)
Vol. 1860: A. Borisyuk, G.B. Ermentrout, A. Friedman, D. Terman, Tutorials in Mathematical Biosciences I. Mathematical Neurosciences (2005)
Vol. 1861: G. Benettin, J. Henrard, S. Kuksin, Hamiltonian Dynamics – Theory and Applications, Cetraro, Italy, 1999. Editor: A. Giorgilli (2005)
Vol. 1862: B. Helffer, F. Nier, Hypoelliptic Estimates and Spectral Theory for Fokker-Planck Operators and Witten Laplacians (2005)
Vol. 1863: H. Führ, Abstract Harmonic Analysis of Continuous Wavelet Transforms (2005)
Vol. 1864: K. Efstathiou, Metamorphoses of Hamiltonian Systems with Symmetries (2005)
Vol. 1865: D. Applebaum, B.V. R. Bhat, J. Kustermans, J. M. Lindsay, Quantum Independent Increment Processes I. From Classical Probability to Quantum Stochastic Calculus. Editors: M. Schürmann, U. Franz (2005)
Vol. 1866: O.E. Barndorff-Nielsen, U. Franz, R. Gohm, B. Kümmerer, S. Thorbjønsen, Quantum Independent Increment Processes II. Structure of Quantum Lévy Processes, Classical Probability, and Physics. Editors: M. Schürmann, U. Franz, (2005)

Recent Reprints and New Editions

LECTURE NOTES IN MATHEMATICS  Springer

Edited by J.-M. Morel, F. Takens, B. Teissier, P.K. Maini

Editorial Policy (for the publication of monographs)

1. Lecture Notes aim to report new developments in all areas of mathematics and their applications - quickly, informally and at a high level. Mathematical texts analysing new developments in modelling and numerical simulation are welcome.

   Monograph manuscripts should be reasonably self-contained and rounded off. Thus they may, and often will, present not only results of the author but also related work by other people. They may be based on specialised lecture courses. Furthermore, the manuscripts should provide sufficient motivation, examples and applications. This clearly distinguishes Lecture Notes from journal articles or technical reports which normally are very concise. Articles intended for a journal but too long to be accepted by most journals, usually do not have this "lecture notes" character. For similar reasons it is unusual for doctoral theses to be accepted for the Lecture Notes series, though habilitation theses may be appropriate.

2. Manuscripts should be submitted either to Springer's mathematics editorial in Heidelberg, or to one of the series editors. In general, manuscripts will be sent out to 2 external referees for evaluation. If a decision cannot yet be reached on the basis of the first 2 reports, further referees may be contacted: The author will be informed of this. A final decision to publish can be made only on the basis of the complete manuscript, however a refereeing process leading to a preliminary decision can be based on a pre-final or incomplete manuscript. The strict minimum amount of material that will be considered should include a detailed outline describing the planned contents of each chapter, a bibliography and several sample chapters.

   Authors should be aware that incomplete or insufficiently close to final manuscripts almost always result in longer refereeing times and nevertheless unclear referees' recommendations, making further refereeing of a final draft necessary.

   Authors should also be aware that parallel submission of their manuscript to another publisher while under consideration for LNM will in general lead to immediate rejection.

3. Manuscripts should in general be submitted in English. Final manuscripts should contain at least 100 pages of mathematical text and should always include

   – a table of contents;
   – an informative introduction, with adequate motivation and perhaps some historical remarks: it should be accessible to a reader not intimately familiar with the topic treated;
   – a subject index: as a rule this is genuinely helpful for the reader.

For evaluation purposes, manuscripts may be submitted in print or electronic form, in the latter case preferably as pdf- or zipped ps-files. Lecture Notes volumes are, as a rule, printed digitally from the authors' files. To ensure best results, authors are asked to use the LaTeX2e style files available from Springer's web-server at:

ftp://ftp.springer.de/pub/tex/latex/svmonot1/ (for monographs).

Additional technical instructions, if necessary, are available on request from: lnm@springer.com.

4. Careful preparation of the manuscripts will help keep production time short besides ensuring satisfactory appearance of the finished book in print and online. After acceptance of the manuscript authors will be asked to prepare the final LaTeX source files (and also the corresponding dvi-, pdf- or zipped ps-file) together with the final printout made from these files. The LaTeX source files are essential for producing the full-text online version of the book (see www.springerlink.com/content/110312 for the existing online volumes of LNM).

   The actual production of a Lecture Notes volume takes approximately 12 weeks.

5. Authors receive a total of 50 free copies of their volume, but no royalties. They are entitled to a discount of 33.3% on the price of Springer books purchased for their personal use, if ordering directly from Springer.

6. Commitment to publish is made by letter of intent rather than by signing a formal contract. Springer-Verlag secures the copyright for each volume. Authors are free to reuse material contained in their LNM volumes in later publications: a brief written (or e-mail) request for formal permission is sufficient.

Addresses:

Professor J.-M. Morel, CMLA,
École Normale Supérieure de Cachan,
61 Avenue du Président Wilson, 94235 Cachan Cedex, France
E-mail: Jean-Michel.Morel@cmla.ens-cachan.fr

Professor F. Takens, Mathematisch Instituut,
Rijksuniversiteit Groningen, Postbus 800,
9700 AV Groningen, The Netherlands
E-mail: F.Takens@math.rug.nl

Professor B. Teissier, Institut Mathématique de Jussieu,
UMR 7586 du CNRS, Équipe "Géométrie et Dynamique",
175 rue du Chevaleret
75013 Paris, France
E-mail: teissier@math.jussieu.fr

For the "Mathematical Biosciences Subseries" of LNM:

Professor P.K. Maini, Center for Mathematical Biology,
Mathematical Institute, 24-29 St Giles,
Oxford OX1 3LP, UK
E-mail: maini@maths.ox.ac.uk

Springer, Mathematics Editorial I, Tiergartenstr. 17
69121 Heidelberg, Germany,
Tel.: +49 (6221) 487-8259
Fax: +49 (6221) 4876-8259
E-mail: lnm@springer.com